Selected Titles in This Series

(Continued in the back of this publication)

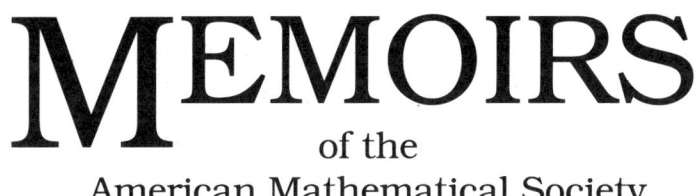

MEMOIRS
of the
American Mathematical Society

Number 586

Crossed Products with Continuous Trace

Siegfried Echterhoff

September 1996 • Volume 123 • Number 586 (first of 4 numbers) • ISSN 0065-9266

American Mathematical Society
Providence, Rhode Island

1991 *Mathematics Subject Classification*.
Primary 46L40; Secondary 22D25, 22D30.

Library of Congress Cataloging-in-Publication Data

Echterhoff, Siegfried, 1960–
    Crossed products with continuous trace / Siegfried Echterhoff.
        p.   cm. – (Memoirs of the American Mathematical Society, ISSN 0065-9266; no. 586)
    "September 1996, volume 123, number 586 (first of 4 numbers)."
    Includes bibliographical references.
    ISBN 0-8218-0563-0 (alk. paper)
    1. C*-algebras.   2. Locally compact groups.   3. Representations of groups.   4. Crossed products.   I. Title.   II. Series.
    QA3.A57   no. 586
    [QA326]
    510 s–dc20
    [512′.55]
                                                                                    96-21893
                                                                                    CIP

Memoirs of the American Mathematical Society

This journal is devoted entirely to research in pure and applied mathematics.

Subscription information. The 1996 subscription begins with Number 568 and consists of six mailings, each containing one or more numbers. Subscription prices for 1996 are $391 list, $313 institutional member. A late charge of 10% of the subscription price will be imposed on orders received from nonmembers after January 1 of the subscription year. Subscribers outside the United States and India must pay a postage surcharge of $25; subscribers in India must pay a postage surcharge of $43. Expedited delivery to destinations in North America $30; elsewhere $92. Each number may be ordered separately; *please specify number* when ordering an individual number. For prices and titles of recently released numbers, see the New Publications sections of the *Notices of the American Mathematical Society*.

Back number information. For back issues see the *AMS Catalog of Publications*.

Subscriptions and orders should be addressed to the American Mathematical Society, P. O. Box 5904, Boston, MA 02206-5904. *All orders must be accompanied by payment.* Other correspondence should be addressed to Box 6248, Providence, RI 02940-6248.

Memoirs of the American Mathematical Society is published bimonthly (each volume consisting usually of more than one number) by the American Mathematical Society at 201 Charles Street, Providence, RI 02904-2213. Periodicals postage paid at Providence, Rhode Island. Postmaster: Send address changes to Memoirs, American Mathematical Society, P. O. Box 6248, Providence, RI 02940-6248.

11/96 umm

Contents

CONTENTS

Abstract

We prove necessary and sufficient conditions for a separable twisted covariant system (A, G, α, τ) to have a continuous-trace crossed product $A \rtimes_{\alpha,\tau} G$ under the general assumptions that A has continuous trace and G/N_τ is abelian. We give a complete description of such systems if the G-quasi-orbit space of $\widehat{A}$ is Hausdorff or if the stability groups for the action of G on $\widehat{A}$ vary continuously. We also prove some relations for the Dixmier-Douady class $\delta(A \rtimes_{\alpha,\tau} G)$ in terms of the Dixmier-Douady class $\delta(A)$ of A and the twisted action (α, τ) of G on A. On our way to do this we introduce the notion of choices of maximally pointwise unitary subgroups for (A, G, α, τ) and give necessary and sufficient conditions for $A \rtimes_{\alpha,\tau} G$ to be of type I in terms of such choices. In a final chapter we apply our results to the study of group C^*-algebras of some solvable and nilpotent Lie groups.

Key words and phrases. C^*-algebras, C^*-dynamical systems, twisted covariant systems, twisted crossed products, continuous trace, duality, induced representations, regularizations, type I crossed products, group C^*-algebras, nilpotent groups, solvable groups.

This work constitutes the author's Habilitationsschrift submitted to the University of Paderborn in November 1993

Received by the editor August 5, 1994

Introduction

During the last fifteen years the study of covariant systems (or C^*-dynamical systems) (A, G, α) and their crossed products $A \rtimes_\alpha G$ has grown to its own subject in the theory of operator algebras. In this work we are mainly interested in the study of separable abelian covariant systems, i.e. covariant systems (A, G, α) where A is a separable C^*-algebra and G is a second countable abelian locally compact group, such that A and $A \rtimes_\alpha G$ have continuous trace. The main motivation for this study is the fact that the knowledge of such systems provides important information about the structure of the group C^*-algebras of many solvable and nilpotent groups (see [65] for an excellent discussion of this theory).

If A is a C^*-algebra, then let $\widehat{A}$ denote the set of all equivalence classes of irreducible $*$-representations of A, the *dual space* of A, equipped with the Jacobson topology. An element a of the positive cone A^+ of A is called a *continuous trace element of A* if the map

$$\widehat{A} \to \mathbb{R} \cup \{\infty\}; \rho \mapsto \operatorname{Tr} \rho(a)$$

is finite and continuous on $\widehat{A}$, where $\operatorname{Tr} \rho(a)$ denotes the usual trace of the operator $\rho(a)$ on the Hilbert space $\mathcal{H}_\rho$ of ρ. A is said to be a C^*-*algebra with continuous trace* if the linear hull $\mathfrak{c}_A$ of all continuous trace elements of A forms a dense ideal of A.

C^*-algebras with continuous trace are very well understood by their representation theory. If A has continuous trace, then $\widehat{A}$ is a locally compact Hausdorff space and A may be realized as a section algebra $\Gamma_0(E)$ of a C^*-bundle $p : E \to \widehat{A}$ such that each fiber A_ρ is isomorphic to the compact operators on the Hilbert space $\mathcal{H}_\rho$. If, in addition, A is separable and stable, i.e., if $A \cong A \otimes \mathcal{K}$, where $\mathcal{K}$ denotes the algebra of compact operators on $l^2(\mathbb{N})$, then A is a section algebra of a locally trivial C^*-bundle. This implies that each point in $\widehat{A}$ has an open neighborhood W such that $I_W = \bigcap\{\ker \sigma; \sigma \in \widehat{A} \smallsetminus W\}$ is isomorphic to $C_0(W, \mathcal{K})$. Moreover, associated to each separable continuous trace C^*-algebra A, there is an element $\delta(A)$ of the third Čech cohomology group $H^3(\widehat{A}, \mathbb{Z})$, the so-called *Dixmier-Douady class* of A. If A and B are two separable and stable continuous trace C^*-algebras with $\widehat{A} = \widehat{B} = \Omega$, then $\delta(A) = \delta(B)$ if and only if there exists an isomorphism between A and B which induces the identity map on Ω (for good expositions of this theory see for instance [9, 66, 59]).

Crossed products with continuous trace were first investigated by Green in [32], where he examined transformation group algebras $C_0(\Omega) \rtimes G$ of locally compact transformation groups (G, Ω). If G and Ω are second countable, and if G acts freely on Ω, i.e., if all stability groups are trivial, then he showed that $C_0(\Omega) \rtimes G$ has continuous trace if and only if G acts *properly* on Ω, which means that the map

$$p : \Omega \times G \to \Omega \times \Omega; (x, s) \mapsto (x, sx)$$

is proper in the sense that inverse images of compact sets are compact. Green's result was extended to non-free actions by Williams in [70]. Apart from some related results for actions of non-abelian groups, he showed that, if G is abelian, $C_0(\Omega) \rtimes G$ has continuous trace if and only if the stabilizer map $S : \Omega \to \mathfrak{K}(G); x \mapsto S_x$ is continuous and G acts σ-properly on Ω in the sense that the map

$$p : \Omega \times_S G \to \Omega \times \Omega; (x, \dot{s}) \mapsto (x, sx)$$

is proper. Here $\mathfrak{K}(G)$ denotes the set of all closed subgroups of G equipped with Fell's topology [21], and $\Omega \times_S G$ denotes the quotient of $\Omega \times G$ by the equivalence relation $\sim_S$ defined by

$$(x, s) \sim_S (y, t) \Leftrightarrow x = y \text{ and } s \in tS_x.$$

If the stabilizer map is constant, say $S_x = S$ for all $x \in \Omega$, then the latter condition means that the action of G factors through a free and proper action of G/S on Ω.

Although we just saw that there exist quite satisfactory theorems about transformation group algebras with continuous trace, there have been only partial results towards a description of general systems (A, G, α) with continuous trace crossed products $A \rtimes_\alpha G$, even when G is abelian and A has continuous trace (which will always be assumed in what follows). However, there are two extreme cases which are known quite well so far: If the natural action of G on $\widehat{A}$ given by $(s, \rho) \to \rho \circ \alpha_{s^{-1}}$ is free, then it was shown by Raeburn and Rosenberg [54, Theorem 1.1] that the crossed product $A \rtimes_\alpha G$ has continuous trace whenever G acts properly on $\widehat{A}$. More recently, it was pointed out by Olesen and Raeburn [46, Theorem 3.1] that in the case of a free action on $\widehat{A}$ the properness of this action is also necessary for $A \rtimes_\alpha G$ to have continuous trace, thus obtaining an analogue of Green's result about free transformation groups as cited above.

The other extreme situation which is well understood is the case where G acts trivially on $\widehat{A}$ and where all Mackey obstructions of (A, G, α) vanish, which means that each $\rho \in \widehat{A}$ may be extended to a covariant representation of (A, G, α). If (A, G, α) is such a system, then α is called *pointwise unitary*. If

α is a pointwise unitary action, then Raeburn and Olesen showed in [**46**, Theorem 1.10] that $A \rtimes_\alpha G$ has continuous trace and $(A \rtimes_\alpha G)\hat{\ }$ is a free and proper $\widehat{G}$-space with respect to the dual action $\widehat{\alpha}$ of $\widehat{G}$ on $A \rtimes_\alpha G$.

Putting these cases together it is also possible to deal with separable abelian systems (A, G, α) which have a constant stabilizer S such that S acts pointwise unitarily on A, or, more generally, systems with constant stabilizer and constant Mackey obstruction (see [**19**, Theorem 6]). It turns out that, even if the constant stabilizer S acts pointwise unitarily on A, the answer to the question whether $A \rtimes_\alpha G$ has continuous trace differs from the answer for commutative A: In general, $A \rtimes_\alpha G$ has continuous trace if and only if the natural action of G/S on $(A \rtimes_\alpha S)\hat{\ }$ is proper. If A is commutative the latter condition is equivalent to saying that G/S acts properly on $\widehat{A}$, but for arbitrary continuous trace algebras A this is not true (see for instance [**19**, Example 1]).

Up to now there has been no satisfactory result describing systems (A, G, α) which have continuous trace crossed products $A \rtimes_\alpha G$ when A is not commutative and the stabilizers and Mackey obstructions of the system are allowed to vary. It is the aim of this treatise to fill this gap: If (A, G, α) is a separable abelian system such that A has continuous trace, then we give necessary and sufficient conditions for $A \rtimes_\alpha G$ to have continuous trace under the additional assumption that (A, G, α) satisfies one of the following two conditions:

(1) The G-quasi-orbit space $\mathcal{Q}_G(\widehat{A})$ of $\widehat{A}$ is Hausdorff.

(2) The stabilizer map $S : \widehat{A} \to \mathfrak{K}(G); \rho \mapsto S_\rho$ is continuous.

Recall that the quasi-orbit space $\mathcal{Q}_G(M)$ of a topological G-space M is the quotient space of M by the equivalence relation $m \sim m' \Leftrightarrow \overline{G(m)} = \overline{G(m')}$, where $G(m) = \{sm; s \in G\}$ denotes the G-orbit of m. Note that by [**19**, Example 2] and Example 6.3.1 below, and unlike the transformation group case, both conditions given above are in general not necessary for $A \rtimes_\alpha G$ to have continuous trace. However, in many interesting situations (for instance if G is compact or discrete or all Mackey obstructions of (A, G, α) vanish) at least one of these conditions turns out to be indispensable for $A \rtimes_\alpha G$ to have continuous trace, so that our results give a description of covariant systems with continuous trace crossed products in a very broad generality.

We are now going to state our results in more detail. Although almost all our results are proved in the more general situation of abelian twisted covariant systems (A, G, α, τ) in the sense of Green [**33**], we restrict here to the setting of ordinary systems (A, G, α). Let us first recall the definition of the Mackey obstructions of a separable covariant system (A, G, α) with A being type I. For this let $\rho \in \widehat{A}$ and let S_ρ denote the stabilizer of ρ. We say that S_ρ acts *pointwise unitarily* at ρ, if there exists a unitary representation U of S_ρ which extends ρ to a covariant representation (ρ, U) of (A, S_ρ, α). It is not possible to find such an extension in general, but there always exist a so-called *multiplier* $\omega_\rho \in Z^2(S_\rho, \mathbb{T})$ (Moore cohomology) and an ω_ρ-*representation*, say L, of S_ρ on $\mathcal{H}_\rho$ which extends ρ to an ω_ρ-covariant representation (ρ, L) of (A, S_ρ, α). The unique class of ω_ρ in the second Moore cohomology group $H^2(S_\rho, \mathbb{T})$ is called the *Mackey obstruction* (for extending ρ to its stabilizer) of (A, G, α) at ρ. Of course, the Mackey obstruction vanishes at ρ if and only if S_ρ acts pointwise unitarily at ρ.

If all Mackey obstructions vanish, then α is said to be pointwise unitary on the stabilizers, and it is natural to ask whether the description of crossed products with continuous trace by actions which are pointwise unitary on a fixed stabilizer, as given above, may be extended to actions which are pointwise unitary on continuously varying stabilizers. In order to state a similar result one has to replace the crossed product $A \rtimes_\alpha S$ by a so-called *subgroup crossed product* $A \rtimes_\alpha \Omega^S$ as defined in [**56, 13**] (basing on very early constructions given by Glimm and Fell in [**27, 24**]). This subgroup crossed product may be thought of as a "fiber C^*-algebra" with fibers $A_\rho \rtimes_\alpha S_\rho$. There is a canonical action, say γ^S, of G on $A \rtimes_\alpha \Omega^S$, and the natural extension of the description of systems (A, G, α) with continuous trace crossed products $A \rtimes_\alpha G$ for actions which are pointwise unitary on the constant stabilizer is

THEOREM 1. *Suppose that (A, G, α) is a separable abelian system such that A has continuous trace and α is pointwise unitary on the continuously varying stabilizers. Then $A \rtimes_\alpha G$ has continuous trace if and only if G acts σ-properly on $(A \rtimes_\alpha \Omega^S)\widehat{\ }$.*

Although, this result provides good progress, it does not give any information about systems with non-vanishing Mackey obstructions. In order to overcome this problem, we introduce the notion of choices of maximally pointwise unitary subgroups for (A, G, α) as follows: Suppose that (A, G, α) is a separable abelian system such that A is type I. If $\rho \in \widehat{A}$, then a closed subgroup L of G is called *maximally ρ-unitary* if L is a maximal subgroup of G with respect to the property that ρ may be extended to a covariant representation, say (ρ, V), of (A, L, α). A map $H : \widehat{A} \to \mathfrak{K}(G); \rho \mapsto H_\rho$ is called a *choice of maximally pointwise unitary subgroups for (A, G, α)*, if H is constant on quasi-orbits in $\widehat{A}$, and if for each $\rho \in \widehat{A}$, H_ρ is a maximally ρ-unitary subgroup of G. We show in Chapter 3 that choices of maximally pointwise unitary subgroups always exist, and that they play a similar role in the representation theory of the crossed product $A \rtimes_\alpha G$ as is played by the stability groups via the Mackey-Green machine. We will see that every primitive ideal of $A \rtimes_\alpha G$ is equal to the kernel of some induced representation $\mathrm{ind}_{H_\rho}^G(\rho \times V)$, where (ρ, V) is an extension of ρ to (A, H_ρ, α). If $A \rtimes_\alpha G$ is type I, then it even turns out that $\mathrm{ind}_{H_\rho}^G(\rho \times V)$ is irreducible, so that all elements in $(A \rtimes_\alpha G)\widehat{\ }$ are induced from some H_ρ. The big advantage of working with choices of maximally pointwise unitary subgroups is the fact that no Mackey obstructions appear for the systems (A_ρ, H_ρ, α). In particular, all Mackey obstructions of (A, G, α) vanish exactly when the stabilizer map $S : \widehat{A} \to \mathfrak{K}(G); \rho \mapsto S_\rho$ is the only choice of maximally pointwise unitary subgroups for (A, G, α). Moreover, if $H : \widehat{A} \to \mathfrak{K}(G)$ is a *continuous* choice of maximally pointwise unitary subgroups for (A, G, α), then we can also form a subgroup crossed product $A \rtimes_\alpha \Omega^H$ with fibers $A_\rho \rtimes_\alpha H_\rho$, together with a canonical action of G on this algebra. Thus, Theorem 1 stated above will be obtained as a special case of the following theorem (see Theorem 5.3.2 below).

THEOREM 2. *Let (A, G, α) be a separable abelian system such that A has continuous trace. Suppose that there is a continuous choice $H : \widehat{A} \to \mathfrak{K}(G); \rho \mapsto H_\rho$ of maximally pointwise unitary subgroups for (A, G, α). Then $A \rtimes_\alpha G$ has continuous trace if and only if G acts σ-properly on $(A \rtimes_\alpha \Omega^H)\widehat{}$.*

In the setting of the above theorem we are also able to deduce a relation between the Dixmier-Douady classes of A and $A \rtimes_\alpha G$.

After having Theorem 2 it is natural to ask whether the existence of a continuous choice of maximally pointwise unitary subgroups is necessary for $A \rtimes_\alpha G$ to have continuous trace. The answer is *yes* if α is pointwise unitary on the stabilizers, but *no* in general. However, it turns out that a necessary condition for $A \rtimes_\alpha G$ to have continuous trace is that maximal pointwise unitary subgroups vary continuously in the sense that, whenever $\rho_n \to \rho$ in $\widehat{A}$ and $(H_n)_{n \in \mathbb{N}}$ is a sequence of subgroups of G such that H_n is maximally ρ_n-unitary for all $n \in \mathbb{N}$, then $H_n \to H$ in $\mathfrak{K}(G)$ implies that H is a maximally ρ-unitary subgroup of G. Note that this does *not* imply the existence of a continuous choice of maximally pointwise unitary subgroups (see Example 6.3.1 below). Yet, we obtain the following result (see Theorem 5.4.3)

THEOREM 3. *Let (A, G, α) be a separable abelian system such that A has continuous trace and such that the quasi-orbit space $\mathcal{Q}_G(\widehat{A})$ of $\widehat{A}$ is Hausdorff. Then $A \rtimes_\alpha G$ has continuous trace if and only if the following conditions are satisfied:*

(1) *(A, G, α) has continuously varying maximally pointwise unitary subgroups.*

(2) *If Λ is any closed G-invariant subset of $\widehat{A}$ such that there exists a continuous choice $H : \Lambda \to \mathfrak{K}(G)$ of maximally pointwise unitary subgroups for (A_Λ, G, α) (with $A_\Lambda = A/\ker \Lambda$), then G acts σ-properly on $(A_\Lambda \rtimes_\alpha \Lambda^H)\widehat{}$.*

Moreover, if $\mathcal{Q}_G(\widehat{A})$ is not assumed to be Hausdorff, then Conditions (1) and (2) are still necessary for $A \rtimes_\alpha G$ to have continuous trace.

The Hausdorff assumption on the quasi-orbit space $\mathcal{Q}_G(\widehat{A})$ in Theorem 3 was required for constructing continuous choices of maximally pointwise unitary subgroups on G-invariant closed subsets of $\widehat{A}$ which are big enough to imply that $A \rtimes_\alpha G$ has continuous trace. We do not know whether this direction of Theorem 3 remains true if we drop the Hausdorff assumption. However, if G is compact or discrete, then it turns out that the quasi-orbit space has to be Hausdorff if $A \rtimes_\alpha G$ has continuous trace. Moreover, if the Mackey obstructions of (A, G, α)

vanish, then Condition (1) just says that (A, G, α) has continuously varying stabilizers, and we are back to the situation of Theorem 1.

We continue by looking more closely at abelian systems (A, G, α) which have continuously varying stabilizers. Here we can prove necessary and sufficient conditions for $A \rtimes_\alpha G$ to be a C^*-algebra with continuous trace which, in many situations, are easier to handle than those given before. If $A \cong \mathcal{K}$, the algebra of compact operators, it is well-known that $\mathcal{K} \rtimes_\alpha G$ has continuous trace if and only if the Mackey obstruction $[\omega]$ of $(\mathcal{K}, G, \alpha)$ is type I (for a very extensive discussion of such systems see [19, §1]). By a classical result of Baggett and Kleppner [3] this is true exactly when the homomorphism $h_\omega : G \to \widehat{G}$ defined by $h_\omega(s)(t) = \omega(s,t)\omega(t,s)^{-1}$ has closed range and is open as a map onto its image. If this is true, then h_ω factors through an isomorphism between G/Σ and $\widehat{G/\Sigma}$, where $\Sigma = \ker h_\omega$ denotes the *symmetry group* of $[\omega]$.

For general systems (A, G, α) with continuous stabilizer map $S : \widehat{A} \to \mathfrak{K}(G)$ we obtain maps $h_{\omega_\rho} : S_\rho \to \widehat{S}_\rho$, where $[\omega_\rho]$ is the Mackey obstruction of (A, G, α) at ρ. If we put $\Omega = \widehat{A}$, then we can construct a map

$$h^\alpha : \Omega^S \to \Omega \times_{S^\perp} \widehat{G}; h^\alpha(x, s) = (x, h_{\omega_x}(s)).$$

Here, as before, $\Omega \times_{S^\perp} \widehat{G}$ denotes the quotient of $\Omega \times \widehat{G}$ by the equivalence relation $(x, \chi) \sim_{S^\perp} (y, \mu) \Leftrightarrow x = y$ and $\chi \in \mu S_x^\perp$. It turns out that h^α is always continuous. In analogy to actions on $\mathcal{K}$ we obtain (see Theorem 5.5.2 below)

THEOREM 4. *Let (A, G, α) be a separable abelian system such that A has continuous trace and such that the stabilizer map $S : \Omega = \widehat{A} \to \mathfrak{K}(G)$ is continuous. Then $A \rtimes_\alpha G$ has continuous trace if and only if the following conditions are satisfied:*
  (1) *The map $h^\alpha : \Omega^S \to \Omega \times_{S^\perp} \widehat{G}$ has closed range and is open as a map onto its image.*
  (2) *$(A \rtimes_\alpha \Omega^S)\widehat{}$ is a σ-proper G-space.*

If $\widehat{A}$ itself is a σ-proper G-space, then it is not hard to see that G acts automatically σ-properly on $(A \rtimes_\alpha \Omega^S)\widehat{}$ provided all Mackey obstructions $[\omega_\rho]$ are type I and the symmetrizer map $\Sigma : \widehat{A} \to \mathfrak{K}(G); \rho \to \Sigma_\rho = \ker h_{\omega_\rho}$ is continuous. It turns out that the type I'ness of the Mackey obstructions and the continuity of Σ is already a consequence of Condition (1). Thus, if G acts σ-properly on $\widehat{A}$, then $A \rtimes_\alpha G$ has continuous trace if and only if Condition (1) is satisfied. We conjecture that Condition (1) is equivalent to to the continuity of Σ together with the type I'ness of all Mackey obstructions of (A, G, α). If all stabilizers of (A, G, α) are constantly equal to a fixed subgroup of G which is a compactly generated Lie group, or if G and all stabilizers are vector groups, then we succeeded in showing that this conjecture is true (see Theorems 5.5.13 and 5.5.16 below).

Our paper is organized as follows. We start in Chapter 1 with a detailed exposition of the theory of twisted crossed products and the Mackey-Green machine as developed by Mackey, Takesaki, Rieffel, Green, Sauvageot, Gootman, Rosenberg and others (see [**39, 40, 41, 62, 63, 69, 33, 67, 31**]). One major tool in our proofs will be the notion of Morita equivalent twisted actions, so in Chapter 2 we recall the basic definitions and facts about Morita equivalent twisted actions as presented in [**15**] and [**16**].

We proceed in Chapter 3 by studying the structure of crossed products of separable abelian type I systems (A, G, α) in terms of choices of maximally pointwise unitary subgroups and give necessary and sufficient conditions for $A \rtimes_\alpha G$ to be type I. If both, A and $A \rtimes_\alpha G$ are type I, then we give a precise description of the Mackey obstructions of the dual system $(A \rtimes_\alpha G, \widehat{G}, \widehat{\alpha})$ in terms of the Mackey obstructions of (A, G, α). It might be interesting to mention that one consequence of our results in Chapter 3 is the fact that every two-step solvable separable type I group G (i.e. the group C^*-algebra $C^*(G)$ is type I) is monomial (Theorem 3.2.3), which means that every element in $\widehat{G}$ is induced from some one-dimensional representation. This extends the well-known result that every separable type I nilpotent group is monomial (see [**34, 5, 2**] for the study of monomial locally compact groups).

Subgroup actions, subgroup crossed products and other subgroup algebras are introduced in Chapter 4, following constructions given in [**27, 24, 56, 12, 13**]. In the same chapter we prove a number of important technical results which are needed later in this work.

All theorems about crossed products with continuous trace are stated and proved in Chapter 5. One key for obtaining these results is a version of Raeburn's and Olesen's theorem [**46**, Theorem 1.10] (see the discussions above) for pointwise unitary actions of continuously varying subgroups on continuous trace algebras. If Ω^H acts pointwise unitarily on A in the sense that the Mackey obstructions of the fiber-systems (A_ρ, H_ρ, α) vanish, then, as a natural extension of [**46**, Theorem 1.10], we show that $(A \rtimes_\alpha \Omega^H)\widehat{}$ is a σ-proper $\widehat{G}$-space with respect to a canonical *dual* $\widehat{G}$-action on $A \rtimes_\alpha \Omega^H$, and that $A \rtimes_\alpha \Omega^H$ is isomorphic to the pull-back of A via the canonical projection from $(A \rtimes_\alpha \Omega^H)\widehat{} \to \widehat{A}$ (Theorem 5.2.9).

Let us explain in a few words the differences between the proof of [**46**, Theorem 1.10] and our proof in the more general setting of subgroup actions. The proof of Olesen and Raeburn bases on a result of Rosenberg, which says that pointwise unitary actions of compactly generated abelian groups are automatically locally unitary. This means that each $\rho \in \widehat{A}$ has an open neighborhood W such that the action of G on $I_W = \bigcap\{\ker \sigma; \sigma \in \widehat{A} \smallsetminus W\}$ is implemented by a strictly continuous homomorphism u from G into the unitaries $\mathcal{U}(I_W)$ of the multiplier algebra $\mathcal{M}(I_W)$ of I_W. It follows that each element of $(A \rtimes_\alpha G)\widehat{}$ is contained in the dual space of an ideal $I_W \rtimes_\alpha G$, which in turn is isomorphic to $I_W \otimes C_0(\widehat{G})$. This implies that $A \rtimes_\alpha G$ has continuous trace and that

$(A \rtimes_\alpha G)\widehat{\,}$ is a locally trivial $\widehat{G}$-bundle whenever α is a pointwise unitary action of a compactly generated group G on a continuous trace algebra A (this situation was extensively discussed in [**52**] and [**54**]). The proof of [**46**, Theorem 1.10] is based on this result by decomposing $A \rtimes_\alpha G$ by a compactly generated open subgroup H of G in order to split the problem into a compactly generated step H (as described above) and a discrete step G/H (which uses the compactness of $\widehat{G/H}$).

Unfortunately, this method of proof does not work in case of continuously varying subgroups, since it is not possible to find any nice class of groups such that pointwise unitary subgroup actions of groups in this class are automatically locally unitary (in a generalized sense). In fact [**56**, Example 4.7] shows that there exists an action of $\mathbb{R}$ on $\mathbb{C}^2$ which is σ-proper but not locally σ-trivial (i.e. which has no local sections), which implies that the dual action of $\mathbb{R}$ on the continuous trace algebra $C_0(\mathbb{C}^2) \rtimes \mathbb{R}$ is pointwise unitary on the continuously varying stabilizers but not locally unitary on the stabilizers. We overcome these problems by observing that, although σ-properness and pointwise unitariness does not imply local σ-triviality or local unitariness, it does imply σ-triviality or unitariness on *closures of subsequences of convergent sequences* and vice versa (for the precise statements see Sections 5.1 and 5.2). It turns out that this is sufficient for obtaining the desired results.

In the first section of Chapter 6 we use the techniques and results of the earlier chapters in order to describe the topological structure of the primitive ideal space $\mathrm{Prim}(A \rtimes_\alpha G)$ of a covariant system (A, G, α) in case where A has continuous trace, G acts σ-properly on $\widehat{A}$, and the symmetrizer map $\Sigma : \widehat{A} \to \mathfrak{K}(G)$ is continuous. Under these assumptions it turns out that $\mathrm{Prim}(A \rtimes_\alpha G)$ is always a σ-proper $\widehat{G}$-space. This extends theorems about the structure of crossed products by actions which are pointwise unitary on a constant stabilizer as given in [**54**, **46**] (see [**57**, **58**, **59**, **60**] for a very sophisticated study of the structure of such crossed products) and gives a very general version of [**56**, Theorem 6.3], where Raeburn and Williams investigated actions which are locally unitary on continuously varying stabilizers. We then show that our methods can be used to give an alternative proof of Rosenberg's theorem (see [**64**, Corollary 2.2]) which says that pointwise unitary actions of abelian compactly generated groups on continuous-trace C^*-algebras are automatically locally unitary. The advantage of our proof is that it does not use Moore cohomology. However, we should mention that [**64**, Corollary 2.2] also gives information about actions of certain non-abelian groups, while our result only applies to abelian groups.

In Section 6.2 we investigate crossed products by actions of the reals $\mathbb{R}$, the integers $\mathbb{Z}$, and the one-dimensional torus group $\mathbb{T}$. If G is one of those groups, then the Mackey obstructions of the system (A, G, α) vanish, since $H^2(H, \mathbb{T}) = \{0\}$ for any subgroup H of G. This, together with some other specialties of these groups, allows to prove that $A \rtimes_\alpha G$ has continuous trace if and only if G acts σ-properly on $\widehat{A}$. Thus for actions of $\mathbb{R}, \mathbb{Z}$ or $\mathbb{T}$ we have exactly the same conditions for $A \rtimes_\alpha G$ to have continuous trace as were proved by Williams in the transformation group case [**70**]. Finally, in Section 6.3, we produce examples which illustrate some subtleties of the theory of crossed products with continuous trace. There we also demonstrate how our results can be applied for the study of group C^*-algebras of certain nilpotent and solvable groups.

It is a pleasure to thank Eberhard Kaniuth, Iain Raeburn, Jonathan Rosenberg and Dana Williams for their strong support and for various interesting discussions on the subject. I'm especially grateful to Dana Williams for providing Proposition 5.4.4 below and for reading a preliminary version of the manuscript. I should remark that his proposition serves as a key for several important results obtained in Chapter 5.

CHAPTER 1

Preliminaries and basic definitions

In this chapter we recall the basic definitions of twisted covariant systems as introduced by Green in [**33**] and give an outline of the Mackey-Green machine for describing the primitive ideal spaces of twisted crossed products $A \rtimes_{\alpha,\tau} G$ in terms of induced representations.

1.1. Twisted covariant systems and crossed products

A *twisted covariant system* (or twisted C^*-dynamical system) (A, G, α, τ) consists of a *covariant system* (A, G, α), where A is a C^*-algebra, G is a locally compact group and $\alpha : G \to \mathrm{Aut}(A)$ is a strongly continuous homomorphism into the group $\mathrm{Aut}(A)$ of $*$-automorphisms of A, together with a *twisting map* $\tau : N_\tau \to \mathcal{U}(A)$, which is a strictly continuous homomorphism from a closed normal subgroup N_τ of G into the group of unitaries $\mathcal{U}(A)$ of the multiplier algebra $\mathcal{M}(A)$ of A satisfying

$$\tau_n a \tau_{n^{-1}} = \alpha_n(a) \text{ and } \alpha_s(\tau_n) = \tau_{sns^{-1}}$$

for all $n \in N_\tau$, $s \in G$ and $a \in A$. The pair (α, τ) is called a *twisted action* of G on A.

A *covariant representation* of a twisted covariant system (A, G, α, τ) is a pair (π, U), where π is a nondegenerate $*$-representation of A on a Hilbert space $\mathcal{H}_\pi$ and U is a unitary representation of G on the same Hilbert space such that

$$U_s \pi(a) U_s^* = \pi(\alpha_s(a)) \text{ for all } s \in G, a \in A$$

and such that (π, U) *preserves* τ in the sense that

$$\pi(\tau_n) = U_n \text{ for all } n \in N_\tau.$$

If (A, G, α, τ) is a twisted covariant system, then $C_c(G, A, \tau)$ denotes the set of all continuous A-valued functions f on G which satisfy

$$f(ns) = f(s)\tau_{n^{-1}} \text{ for all } s \in G \text{ and } n \in N_\tau,$$

and which have compact support in $\tilde{G} = G/N_\tau$. We define convolution, involution and norm on $C_c(G, A, \tau)$ by

$$f * g(s) = \int_{\tilde{G}} f(t)\alpha_t(g(t^{-1}s)) \, d\tilde{t},$$

$$f^*(s) = \Delta_{\tilde{G}}(\tilde{s}^{-1})\alpha_s(f(s^{-1})^*)$$

and

$$\|f\|_1 = \int_{\tilde{G}} \|f(s)\| \, d\tilde{s},$$

where $\Delta_{\tilde{G}}$ denotes the modular function on $\tilde{G}$. Let $L^1(G, A, \tau)$ denote the completion of $C_c(G, A, \tau)$ by $\| \cdot \|_1$. The *twisted crossed product* $A \rtimes_{\alpha, \tau} G$ of (A, G, α, τ) can be defined as the enveloping C^*-algebra of $L^1(G, A, \tau)$.

There is a one-to-one correspondence between the nondegenerate $*$-representations of $A \rtimes_{\alpha, \tau} G$ and the covariant representations of (A, G, α, τ), which is given as follows: If (π, U) is a covariant representation of (A, G, α, τ) then the corresponding representation of $A \rtimes_{\alpha, \tau} G$, denoted $\pi \times U$, is given by

$$\pi \times U(f) = \int_{\tilde{G}} \pi(f(s))U_s \, d\tilde{s}, \quad f \in L^1(G, A, \tau).$$

$\pi \times U$ is called the *integrated form of* (π, U). For the opposite direction assume that R is a (nondegenerate) representation of $A \rtimes_{\alpha, \tau} G$. If we define $\pi = R \circ i_A$ and $U = R \circ i_G$, where i_A and i_G denote the canonical embeddings of G and A into the multiplier algebra $\mathcal{M}(A \rtimes_{\alpha, \tau} G)$ given by

$$(i_A(a)f)(s) = af(s) \quad \text{and} \quad (i_G(t)f)(s) = \alpha_t(f(t^{-1}s)),$$

$f \in C_c(G, A, \tau)$, then (π, U) is a covariant representation of (A, G, α, τ) with $R = \pi \times U$.

If τ is trivial, i.e., if $N_\tau = \{e\}$, then the procedure above reduces to the usual construction of the *crossed product* $A \rtimes_\alpha G$ of the covariant system (A, G, α) as a completion of $C_c(G, A)$. Note that $A \rtimes_{\alpha, \tau} G$ is always a quotient of the crossed product $A \rtimes_\alpha G$, where the quotient map is given by the extension of the map

$$\Phi : C_c(G, A) \to C_c(G, A, \tau); (\Phi f)(s) = \int_{N_\tau} f(sn)\tau_{sns^{-1}} \, dn$$

to $A \rtimes_\alpha G$. Here Haar measures on G, N_τ and $\tilde{G} = G/N_\tau$ are chosen such that

$$\int_G g(s) \, ds = \int_{\tilde{G}} \int_{N_\tau} g(sn) \, dn \, d\tilde{s}$$

for all $g \in C_c(G)$. Note that the kernel of Φ is just the ideal I_τ which is the intersection of all kernels of representations $\pi \times U$ of $A \rtimes_\alpha G$ such that (π, U) preserves τ (for more details see [33]).

The main reason for introducing twisted covariant systems by Green [33] and others is the possibility of splitting a twisted crossed products $A \rtimes_{\alpha, \tau} G$ by a closed normal subgroup M of G as described in the following example.

Example 1.1.1. Suppose that (A, G, α, τ) is a twisted covariant system and M is a closed normal subgroup of G containing N_τ. Then there is a canonical action, say γ^M, of G on $A \rtimes_{\alpha, \tau} M$, which is given on $f \in C_c(M, A, \tau)$ by

$$\gamma_s^M(f(m)) = \delta(s)\alpha_s(f(s^{-1}ms)),$$

where $\delta(s) = \Delta_{G/N_\tau}(s)\Delta_{G/M}(s^{-1})$ for all $s \in G$. There is also a canonical twisting map, say τ^M, for $(A \rtimes_{\alpha, \tau} M, G, \gamma^M)$, which is given by the canonical embedding of M into $\mathcal{U}(A \rtimes_{\alpha, \tau} M)$, i.e.

$$\tau_m^M f(l) = \alpha_m(f(m^{-1}l))$$

for $m, l \in M$ and $f \in C_c(M, A, \tau)$. The resulting twisted action (γ^M, τ^M) of G on $A \rtimes_{\alpha, \tau} M$ is called the *decomposition twisted action* of G on $A \rtimes_{\alpha, \tau} M$. If we define a homomorphism $\Psi : C_c(G, A, \tau) \to C_c(G, C_c(M, A, , \tau), \tau^M)$ by

$$(\Psi f(s))(m) = \delta(s)f(ms),$$

then it follows from straightforward arguments that Ψ extends to an isomorphism from $A \rtimes_{\alpha, \tau} G$ onto $(A \rtimes_{\alpha, \tau} M) \rtimes_{\gamma^M, \tau^M} G$ (see [**33**, Proposition 1] and [**12**, Section 4]). Note that Ψ transports a covariant representation (π, U) of (A, G, α, τ) to the covariant representation $(\pi \times U|_M, U)$ of $(A \rtimes_{\alpha, \tau} M, G, \gamma^M, \tau^M)$, where $U|_M$ denotes the restriction of U to M.

In particular, if $A = \mathbb{C}$ and τ is trivial, then we obtain a canonical twisted action (γ^M, τ^M) of G on the group C^*-algebra $C^*(M)$ of M such that $C^*(G)$ is isomorphic to $C^*(M) \rtimes_{\gamma^M, \tau^M} G$, and this isomorphism carries a unitary representation U of G to the representation $(U|_M, U)$ of $(C^*(M), G, \gamma^M, \tau^M)$.

The following example shows that twisted actions and crossed products behave nicely with respect to taking quotients by G-invariant ideals of A.

Example 1.1.2. Suppose that (A, G, α, τ) is a twisted covariant system and let I be a closed G-invariant ideal in A. Then there are canonical twisted actions (α^I, τ^I) and $(\alpha^{A/I}, \tau^{A/I})$ of G on I and A/I, respectively, given by

$$\alpha_s^I(b) = \alpha_s(b), \quad \tau_n^I b = \tau_n b$$

and

$$\alpha_s^{A/I}(a + I) = \alpha_s(a) + I, \quad \tau_n^{A/I}(a + I) = \tau_n a + I$$

for all $b \in I$, $a \in A$, $s \in G$ and $n \in N_\tau$. The natural sequence of maps

$$C_c(G, I, \tau^I) \to C_c(G, A, \tau) \to C_c(G, A/I, \tau^{A/I})$$

given by inclusion and composition with the quotient map $A \to A/I$, respectively, extends to the twisted crossed products and gives rise to a short exact sequence

$$0 \to I \rtimes_{\alpha^I, \tau^I} G \to A \rtimes_{\alpha, \tau} G \to (A/I) \rtimes_{\alpha^{A/I}, \tau^{A/I}} G \to 0.$$

Note that an irreducible representation $\pi \times U \in (A \rtimes_{\alpha, \tau} G)\hat{}$ belongs to the closed subset $((A/I) \rtimes_{\alpha^{A/I}, \tau^{A/I}} G)\hat{}$ if and only if $I \subseteq \ker \pi$.

More generally, if F is any locally closed subset of $\widehat{A}$, i.e. F is open in its closure $\overline{F} \subseteq \widehat{A}$, then taking $J = \ker F = \cap\{\ker \rho; \rho \in F\}$ and $I = \ker(\overline{F} \setminus F) = \cap\{\ker \rho; \rho \in \overline{F} \setminus F\}$ we put $A_F = I/J$. F is canonically homeomorphic to $\widehat{A}_F$. If F is invariant under the action of G on $\widehat{A}$ given by $(s, \rho) \mapsto \rho \circ \alpha_{s^{-1}}$, then I and J are G-invariant, too. Thus we obtain a canonical twisted action (α^F, τ^F) of G on A_F and we may identify $(A_F \rtimes_{\alpha^F, \tau^F} G)\widehat{\,}$ with the locally closed subset of $(A \rtimes_{\alpha, \tau} G)\widehat{\,}$ consisting of all $\pi \times U$ such that π *lives on* F in the sense that $J \subseteq \ker \pi$ but $I \not\subseteq \ker \pi$. Note that we will usually denote (α^F, τ^F) (and also (α^I, τ^I) and $(\alpha^{A/I}, \tau^{A/I})$) simply by (α, τ), if no confusion is possible.

The next example shows that every ordinary action of a quotient $\tilde{G} = G/N$ of G on a C^*-algebra B may be viewed as a twisted action of G on B. This identification is fundamental in this work.

Example 1.1.3. Suppose that N is a closed normal subgroup of G and that β is an action of $\tilde{G} = G/N$ on a C^*-algebra B. Let $q : G \to \tilde{G}$ denote the quotient map. Then, if 1_N denotes the trivial homomorphism from N into $\mathcal{U}(B)$, $(\beta \circ q, 1_N)$ is a twisted action of G on B which is called the *twisted action lifted from* β. It is trivially seen that a covariant representation (π, U) of $(B, G, \beta \circ q)$ preserves 1_N if and only if $U = \tilde{U} \circ q$ for some unitary representation $\tilde{U}$ of $\tilde{G}$. Moreover, $C_c(G, B, 1_N)$ consists of all functions which are constant on N-cosets and have compact support in $\tilde{G}$. Thus we may identify $C_c(G, B, 1_N)$ with $C_c(\tilde{G}, B)$ and it follows that $B \rtimes_{\beta \circ q, 1_N} G = B \rtimes_\beta \tilde{G}$.

The last example of twisted actions we want to present in this section plays an important role in the representation theory of twisted crossed products.

Example 1.1.4. Let ω be a multiplier on the separable locally compact group G, i.e. ω is a Borel map $\omega : G \times G \to \mathbb{T}$ which satisfies the multiplier identities
  (1) $\omega(s, e) = \omega(e, s) = 1$ for all $s \in G$,
  (2) $\omega(st, r)\omega(s, t) = \omega(s, tr)\omega(t, r)$ for all $s, t, r \in G$.
We define multiplication on the set $G^\omega = G \times \mathbb{T}$ by

$$(s, z)(t, w) = (st, \omega(s, t)zw), \quad s, t \in G, z, w \in \mathbb{T}.$$

If we equip $G \times \mathbb{T}$ with the Weyl topology, which is a certain locally compact topology on $G \times \mathbb{T}$ which generates the same Borel sets as the product topology, then G^ω becomes a locally compact group. G^ω is a central extension of G by $\mathbb{T}$ and every central extension of G by $\mathbb{T}$ can be obtained in this way (see for instance [**42**]). Two central extensions G^ω and $G^{\omega'}$ of $\mathbb{T}$ by G are isomorphic as extensions of $\mathbb{T}$ by G if and only if ω is similar to ω' in the sense that there exists a Borel function $f : G \to \mathbb{T}$ satisfying $f(e) = 1$ and

$$\omega(s, t) = \omega'(s, t)f(s)f(t)f(st)^{-1}$$

for $s, t \in G$. Note that the set of multipliers on G forms the group $Z^2(G, \mathbb{T})$ of cocycles in Moore cohomology, and ω and ω' are in the same class of the second Moore cohomology group $H^2(G, \mathbb{T})$ if and only if ω is similar to ω'.

Now let $N_\tau = \mathbb{T} \subseteq G^\omega$ and let $\tau : N_\tau \to \mathbb{T} = \mathcal{U}(\mathbb{C}); \tau_z = z$ be the identity. Then τ is a twisting map for the trivial action id of G^ω on $\mathbb{C}$. The representations of the twisted crossed product $C^*(G, \omega) = \mathbb{C} \rtimes_{\mathrm{id}, \tau} G^\omega$ coincide with the unitary representations of G which are multiples of the identity when restricted to $\mathbb{T}$. These representations are, in turn, in a one-to-one correspondence to the ω-representations of G via the restriction to the set $\{(s, 1) \subseteq G^\omega; s \in G\}$. Recall that a map $L : G \to \mathcal{U}(\mathcal{H})$ is called an ω-*representation* of G if L is weakly Borel, and if

$$L_s L_t = \omega(s, t) L_{st}, \quad \text{for all } s, t \in G.$$

$C^*(G, \omega)$ is called the *twisted group algebra defined by* ω. Note that the isomorphism class of $C^*(G, \omega)$ only depends on the class of ω in $H^2(G, \mathbb{T})$.

1.2. Representation and ideal spaces of C^*-algebras

In addition to $\widehat{A}$ and $\mathrm{Prim}(A)$, we will need to work with some more general representation and ideal spaces of a C^*-algebra A. One of these spaces is the space $\mathcal{I}(A)$ of all closed ideals in A, where a subbasis for the topology is given by the sets $U(I) = \{J \in \mathcal{I}(A); J \cap I \neq \emptyset\}$, I running through all of $\mathcal{I}(A)$. Another space is the space $\mathrm{Rep}(A)$ of all equivalence classes of $*$-representations of A with dimension bounded by a fixed cardinal $\aleph$. The last restriction has to be made in order that $\mathrm{Rep}(A)$ be a set. However, we will assume that $\aleph$ is big enough to guarantee that all representations we are interested in have dimension less than $\aleph$. As for $\widehat{A}$, $\mathrm{Rep}(A)$ carries the inverse image of the topology of $\mathcal{I}(A)$ via the canonical map $\pi \mapsto \ker \pi$. The topologies on $\mathcal{I}(A)$ and $\mathrm{Rep}(A)$ defined above are called the *Fell topologies* on these spaces. Note that the Fell topologies restricted to $\widehat{A}$ and $\mathrm{Prim}(A)$, respectively, coincide with the usual Jacobsen topologies (see [9, Chapter 3] and [22, 24] for more details).

While working with $\mathrm{Rep}(A)$, it is very useful to introduce the notion of weak containment. If R and S are two subsets of $\mathrm{Rep}(A)$, then we say that R is *weakly contained* in S, denoted $R \prec S$, if $\ker R \supseteq \ker S$. Here, for any subset $E \subseteq \mathrm{Rep}(A)$, $\ker E = \cap\{\ker \pi; \pi \in E\}$. If $\ker R = \ker S$, then we say that R is *weakly equivalent* to S, denoted $R \sim S$. Moreover, if $\rho \in \mathrm{Rep}(A)$ and $S \subseteq \mathrm{Rep}(A)$, then we say $\rho \prec S$ or $\rho \sim S$ iff $\{\rho\} \prec S$ or $\{\rho\} \sim S$, respectively. It is clear that, restricted to $\widehat{A}$, weak containment gives the closure relation on $\widehat{A}$. The following proposition shows that there is also a relation between weak containment and convergence in $\mathrm{Rep}(A)$ (see [24, Propositions 1.2 and 1.3]).

PROPOSITION 1.2.1. *Let* $(\pi_i)_{i \in I}$ *be a net in* $\mathrm{Rep}(A)$ *and* $\pi, \rho \in \mathrm{Rep}(A)$.
   (1) $\pi_i \to \pi$ *in* $\mathrm{Rep}(A)$ *if and only if* π *is weakly contained in every subnet of* $(\pi_i)_{i \in I}$.
   (2) *If* $\pi_i \to \pi$ *and* $\rho \prec \pi$, *then* $\pi_i \to \rho$.

The following is [**24**, Proposition 1.4].

PROPOSITION 1.2.2. *If (A, G, α) is a covariant system, then*

$$G \times \mathrm{Rep}(A) \to \mathrm{Rep}(A); (s, \pi) \mapsto \pi \circ \alpha_{s^{-1}}$$

is continuous and defines a continuous action of G on $\mathrm{Rep}(A)$.

The next result is also quite convenient when working with nets in $\mathrm{Rep}(A)$. For separable A it is proved in [**68**], but the same arguments also give the general case.

PROPOSITION 1.2.3. *Let $(\pi_i)_{i \in I}$ be a net in $\mathrm{Rep}(A)$ converging to some $\rho \in \widehat{A}$. Suppose that $\mathcal{D}_i \subseteq \widehat{A}$ such that $\mathcal{D}_i \sim \pi_i$ for all $i \in I$. Then there exists a subnet $(\pi_{i_j})_{j \in J}$ of $(\pi_i)_{i \in I}$ and $\rho_j \in \mathcal{D}_{i_j}$ for all $j \in J$ such that $\rho_j \to \rho$ in $\widehat{A}$.*

1.3. Morita equivalence and imprimitivity bimodules

We are now going to recall the notion of Morita equivalence of C^*-algebras as introduced by Rieffel in [**62**, Definition 6.10]. Recall that a *pre-C^*-algebra* is a normed $*$-algebra whose completion is a C^*-algebra. If B_0 is a pre-C^*-algebra, then by a *right* (resp. *left*) *pre-Hilbert B_0-module* we understand a right (resp. left) B_0-module X_0 equipped with a B_0-valued *pre-inner product* $\langle \cdot, \cdot \rangle_{B_0}$, satisfying

    (1) $\langle \xi, \eta \rangle_{B_0} = \langle \eta, \xi \rangle_{B_0}^*$ and $\langle \xi, \xi \rangle_{B_0} \geq 0$ for all $\xi, \eta \in X$.

    (2) $\langle \xi, \eta b \rangle_{B_0} = \langle \xi, \eta \rangle_{B_0} b$ (resp. $\langle b\xi, \eta \rangle_{B_0} = b \langle \xi, \eta \rangle_{B_0}$) for all $\xi, \eta \in X_0$ and $b \in B_0$.

    (3) The linear hull $\langle X_0, X_0 \rangle_{B_0}$ of $\{ \langle \xi, \eta \rangle_{B_0}; \xi, \eta \in X_0 \}$ is dense in B_0.

If B is a C^*-algebra, then X is called a *right* (resp. *left*) *Hilbert B-module* if there is a B-valued *inner product* on X, which satisfies Conditions (1), (2) and (3) above such that in addition $\langle \xi, \xi \rangle_B > 0$ for all $\xi \neq 0$ and such that X is complete with respect to the norm $\| \xi \| = \| \langle \xi, \xi \rangle_B \|^{1/2}$.

Remark 1.3.1. In the literature, the definition of Hilbert modules often not requires Condition (3) above. Hilbert modules which satisfy this condition are usually called *full*. However, since we always need this condition we put it into the definition of Hilbert modules.

DEFINITION 1.3.2. Suppose that B_0 and C_0 are pre-C^*-algebras. By a $B_0 - C_0$ *pre-imprimitivity bimodule* X_0 we understand a left pre-Hilbert B_0- and right pre-Hilbert C_0-bimodule X_0 satisfying the conditions:

    (1) $\xi \langle \eta, \zeta \rangle_C = \langle \xi, \eta \rangle_B \zeta$ for all $\xi, \eta, \zeta \in X$.

    (2) $\langle b\xi, b\xi \rangle_C \leq \|b\|^2 \langle \xi, \xi \rangle_C$ and $\langle \xi c, \xi c \rangle_B \leq \|c\|^2 \langle \xi, \xi \rangle_B$ for all $b \in B$, $\xi \in X$ and $c \in C$.

If B and C are C^*-algebras, then a $B - C$ *imprimitivity bimodule* (sometimes also called *equivalence bimodule*) is a left Hilbert B- and right Hilbert C-bimodule X, satisfying Conditions (1) and (2) above. Two C^*-algebras B and C are called *Morita equivalent* if there exists a $B - C$ imprimitivity bimodule X.

As a consequence of Conditions (1) and (2) in Definition 1.3.2, it follows that the semi-norms on a B_0-C_0 pre-imprimitivity bimodule X_0 given by the B_0- and C_0-valued inner products coincide [**63**, Proposition 3.1]. This allows to extend all actions and inner products to the completions, say B, X and C, of B_0, X_0 and C_0, respectively. By dividing out all elements with length zero, we may assume that these semi-norms are actually norms so that X becomes a $B - C$ imprimitivity bimodule. The reason for defining pre-imprimitivity bimodules rather than only defining imprimitivity bimodules is that in many important applications it is only possible to describe the actions and inner products on dense subalgebras and submodules.

The notion of Morita equivalence is in fact an equivalence relation: Any C^*-algebra B can be made into a $B - B$ imprimitivity bimodule in a canonical way by defining the B-valued inner products of elements b_1, b_2 of B by $b_1^* b_2$ and $b_1 b_2^*$, respectively. Moreover, if X is a $B - C$ imprimitivity bimodule, then we can make X into a $C - B$ imprimitivity bimodule, say X^*, by taking the same inner products, and by defining the left action of C and the right action of B on X^* by

$$c\xi^* = \overline{\xi c^*} \quad \text{and} \quad \xi^* b = \overline{b^* \xi},$$

where we denote by ξ^* the element ξ of X when viewed as an element of X^*. X^* is called the *adjoint* of X. Finally, if X is a $B - C$ imprimitivity bimodule and Y is a $C - D$ imprimitivity bimodule, then the *balanced tensor product* $X \otimes_C Y$ becomes a $B - D$ imprimitivity bimodule, where $X \otimes_C Y$ is the Hausdorff completion of the algebraic tensor product $X \odot Y$ with respect to the D-valued inner product

$$\langle \xi \otimes \eta, \xi' \otimes \eta' \rangle_D = \langle \langle \xi', \xi \rangle_C \eta, \eta' \rangle_D.$$

The B-valued inner product on $X \otimes_C Y$ is defined similarly.

If X is a $B - C$ imprimitivity bimodule, then X implements an equivalence between the representation theories of B and C in the following way: If π is a nondegenerate $*$-representation of C, then we form the Hilbert space $X \otimes_C \mathcal{H}_\pi$ as the Hilbert space completion of the algebraic tensor product $X \odot \mathcal{H}_\pi$ by the pre-inner product given on elementary tensors by

$$\langle \xi \otimes v, \xi' \otimes v' \rangle = \langle \pi(\langle \xi', \xi \rangle_C) v, v' \rangle.$$

The representation $\operatorname{ind}^X \pi$ of B *induced from* π *via* X acts on $X \otimes_C \mathcal{H}_\pi$ by

$$\operatorname{ind}^X \pi(b)(\xi \otimes v) = b\xi \otimes v.$$

Note that $\operatorname{ind}^{X^*}(\operatorname{ind}^X \pi)$ is unitarily equivalent to π, so that inducing via the adjoint bimodule X^* is inverse to inducing via X. By [**62**, Proposition 6.25] we know that inducing via X preserves weak containment. Thus it follows easily that

$$\operatorname{ind}^X : \operatorname{Rep}(C) \to \operatorname{Rep}(B); \pi \mapsto \operatorname{ind}^X \pi$$

is a homeomorphism (by a suitable choice of the bounds for the dimensions of the representations in $\operatorname{Rep}(B)$ and $\operatorname{Rep}(C)$.) Note that $\operatorname{ind}^X \pi$ is irreducible if and only if π is irreducible, which shows that the restriction of ind^X to $\widehat{C}$ implements a homeomorphism between $\widehat{C}$ and $\widehat{B}$.

There is also a direct way to induce ideals from C to B via X. For this assume that I is a closed ideal in C. Then XI becomes a Hilbert I-module in a canonical way and it is not hard to see that the closure I^X of $\langle XI, XI \rangle_B$ is an ideal of B, which is called the *ideal of B induced from I via X*. Note that XI becomes a $I^X - I$ imprimitivity bimodule, while the quotient X/XI can be made canonically into a $B/I^X - C/I$ imprimitivity bimodule by defining the C/I- and B/I^X-valued inner products by $\langle \xi + XI, \eta + XI \rangle_{C/I} = \langle \xi, \eta \rangle_C + I$ and $\langle \xi + XI, \eta + XI \rangle_{B/I^X} = \langle \xi, \eta \rangle + I^X$ (it is shown in the proof of [18, Proposition 2.6] that X/XI is always complete). If $\pi \in \text{Rep}(C)$ such that $I \subseteq \ker \pi$, then we have $I^X \subseteq \text{ind}^X \pi$, and $\text{ind}^X \pi = \text{ind}^{X/XI} \pi$, if π and $\text{ind}^X \pi$ are viewed as representations of C/I and B/I^X, respectively. Moreover, $(\ker \pi)^X = \ker(\text{ind}^X \pi)$ for all $\pi \subset \text{Rep}(C)$ and the map

$$\text{Ind} : \mathcal{I}(C) \to \mathcal{I}(B); I \mapsto I^X$$

is a homeomorphism. All these results can be found in [63, Section 3].

Let us finally mention that if B and C are C^*-algebras with strictly positive elements (for instance if B and C are separable), then B and C are Morita equivalent if and only if they are stably isomorphic, which means that $B \otimes \mathcal{K} \cong C \otimes \mathcal{K}$ where $\mathcal{K}$ denotes the algebra of compact operators on the separable Hilbert space $\mathcal{H}$ (see [4]). Let us further mention that the property of having continuous trace is invariant under Morita equivalence (see [70, Theorem 2.15]). This will be a major tool in our investigation of twisted crossed products with continuous trace.

1.4. Induced representations and the Mackey-Green machine

In this section we want to recall the basic techniques of the Mackey-Green machine for describing the primitive ideal space of a twisted covariant system via induced representations. Assume that (A, G, α, τ) is a twisted covariant system and let H be a closed subgroup of G containing N_τ. Let $(\tilde{\alpha}, \tilde{\tau})$ denote the twisted action of G on $C_0(G/H, A)$ defined by

$$\tilde{\alpha}_s(\varphi)(\dot{t}) = \alpha_s(\varphi(s^{-1}t)) \quad \text{and} \quad (\tilde{\tau}_n \varphi)(\dot{t}) = \tau_n(\varphi(\dot{t}))$$

for all $s \in G, n \in N_\tau$ and $\varphi \in C_0(G/H, A)$, where $\dot{t}$ denotes the left coset space tH of $t \in G$. Let $C_0^\tau = C_c(H, A, \tau)$, $X_0^\tau = C_c(G, A, \tau)$ and $B_0^\tau = C_c(G, C_c(G/H, A), \tilde{\tau})$. If τ is trivial we simply denote these spaces by C_0, X_0 and B_0, respectively. We may regard C_0^τ and B_0^τ as dense subalgebras of the twisted crossed products $C^\tau = A \rtimes_{\alpha, \tau} H$ and $B^\tau = C_0(G/H, A) \rtimes_{\tilde{\alpha}, \tilde{\tau}} G$, respectively, denoted C and B if τ is trivial.

We define left and right actions of B_0^τ and C_0^τ on X_0^τ, respectively, and C_0^τ-

and B_0^τ-valued inner products on X_0^τ by

$$F \cdot \xi(s) = \int_{\tilde{G}} F(t, \dot{s}) \alpha_t(\xi(t^{-1}s)) \, d\tilde{t}$$

$$\xi \cdot f(s) = \int_{\tilde{H}} \xi(sh) \alpha_{sh}(f(h^{-1})) \phi_H(h^{-1}) \, d\tilde{h}$$

$$\langle \xi, \eta \rangle_{B_0^\tau}(s, \dot{r}) = \int_{\tilde{H}} \xi(rh) \alpha_{rh}(\eta^*(h^{-1}r^{-1}s)) \, d\tilde{h}$$

$$\langle \xi, \eta \rangle_{C_0}(h) = \phi_H(h^{-1}) \int_{\tilde{G}} \xi^*(t) \alpha_t(\eta(t^{-1}h)) \, d\tilde{t},$$

where $\phi_H(h) = (\Delta_G(h)/\Delta_H(h))^{1/2}$ for $h \in H$. Here, as usual, $\tilde{G} = G/N_\tau$, $\tilde{H} = H/N_\tau$ and $\tilde{t} = tN_\tau$ for all $t \in G$. Note that ϕ_H disappears whenever H is normal in G. As a slight reformulation of Green's imprimitivity theorem [33, Proposition 3] we obtain

THEOREM 1.4.1 (GREEN). *With the actions and inner products as defined above, X_0^τ may be completed to a $B^\tau - C^\tau$ imprimitivity bimodule.*

PROOF. In case where τ is trivial, the theorem follows directly from [33, Proposition 3], if we keep in mind that we are using left Haar measures instead of the symmetric Haar measures as used in [33]. Let $\Phi^B : B_0 \to B_0^\tau$, $\Phi^C : C_0 \to C_0^\tau$ and $\Phi^X : X_0 \to X_0^\tau$ be given by

$$\Phi^B F(s, \dot{t}) = \int_{N_\tau} F(sn, \dot{t}) \tau_{sns^{-1}} \, dn, \quad \Phi^C f(h) = \int_{N_\tau} f(hn) \tau_{hnh^{-1}} \, dn$$

$$\text{and} \quad \Phi^X \xi(s) = \int_{N_\tau} \xi(sn) \tau_{sns^{-1}} \, dn.$$

Then it is easily seen that these maps transport all actions of B_0 and C_0 and the B_0- and C_0-valued inner products on X_0 onto those on X_0^τ as defined above, and the proof is complete (see also [33, Corollary 5]). $\square$

There is a natural $*$-homomorphism ϱ^H from $A \rtimes_{\alpha,\tau} G$ into $\mathcal{M}(B^\tau)$ which is given by

$$(\varrho^H(f)F)(s, \dot{t}) = \int_{\tilde{G}} f(r) \alpha_r(F(r^{-1}s, r^{-1}t)) \, d\tilde{r}$$

for $f \in C_c(G, A, \tau)$, $F \in C_c(G, C_c(G/H, A), \tilde{\tau})$. If $\rho \times V$ is a representation of $C^\tau = A \rtimes_{\alpha,\tau} H$, then the *induced representation* $\text{ind}_H^G(\rho \times V)$ of $A \rtimes_{\alpha,\tau} G$ is defined by

$$\text{ind}_H^G(\rho \times V) = \text{ind}^{X^\tau}(\rho \times V) \circ \varrho^H,$$

where $\text{ind}^{X^\tau}(\rho \times V)$ is the representation of B^τ induced from $\rho \times V$ via X^τ. Note that by the canonical extension of the left action of B^τ on X^τ to the multiplier algebra $\mathcal{M}(B^\tau)$, ϱ^H becomes the $*$-representation on X^τ given by

$$\varrho^H(f)\xi = f * \xi$$

for all $f \in C_c(G, A, \tau)$ and $\xi \in X_0^\tau$. Thus, for $f \in C_c(G, A, \tau)$ and elementary tensors $\xi \otimes v \in X_0^\tau \odot \mathcal{H}_\rho$ we obtain the formula

$$\mathrm{ind}_H^G(\rho \times V)(f)(\xi \otimes v) = f * \xi \otimes v.$$

Note that inducing $\rho \times V$ to $A \rtimes_{\alpha,\tau} G$ via X^τ as above, or inducing $\rho \times V$ as a representation of the non-twisted crossed product $C = A \rtimes_\alpha H$ to $A \rtimes_\alpha G$ via X gives the same representation, which is a direct consequence of the proof of Theorem 1.4.1. We are now going to state several results about induced representations which are used frequently in this work. The first is the theorem of induction in stages for crossed products (see [**33**, Proposition 8]).

PROPOSITION 1.4.2. *Suppose that (A, G, α, τ) is a twisted covariant system and let $L \supseteq H \supseteq N_\tau$ be closed subgroups of G. Then $\mathrm{ind}_L^G(\mathrm{ind}_H^L(\rho \times V))$ is unitarily equivalent to $\mathrm{ind}_H^G(\rho \times V)$ for all $\rho \times V \in \mathrm{Rep}(A \rtimes_{\alpha,\tau} H)$.*

The next proposition is [**33**, Proposition 7]. It shows that inducing representations behaves well with respect to decomposition of twisted crossed products as defined in Example 1.1.1.

PROPOSITION 1.4.3. *Let (A, G, α, τ) be a twisted covariant system and let $N_\tau \subseteq M \subseteq H$ be closed subgroups of G such that M is normal in G. Let (γ^M, τ^M) be the decomposition twisted action of G on $A \rtimes_{\alpha,\tau} M$, and for $\rho \times V \in \mathrm{Rep}(A \rtimes_{\alpha,\tau} H)$ let $(\rho \times V|_M) \times V$ be the representation of $(A \rtimes_{\alpha,\tau} M) \rtimes_{\gamma^M, \tau^M} H$ corresponding to $\rho \times V$ via the canonical isomorphism between $A \rtimes_{\alpha,\tau} H$ and $(A \rtimes_{\alpha,\tau} M) \rtimes_{\gamma^M, \tau^M} H$. Then $\mathrm{ind}_H^G((\rho \times V|_M) \times V)$ corresponds to $\mathrm{ind}_H^G(\rho \times V)$ via the canonical isomorphism between $A \rtimes_{\alpha,\tau} G$ and $(A \rtimes_{\alpha,\tau} M) \rtimes_{\gamma^M, \tau^M} G$.*

The next proposition is part of [**33**, Proposition 12]. As a consequence we see that inducing representations is compatible with taking quotients of twisted crossed products as in Example 1.1.2.

PROPOSITION 1.4.4. *Let (A, G, α, τ) be a twisted covariant system, H a closed normal subgroup of G containing N_τ and let I be a closed G-invariant ideal in A. Let X^τ, X^{τ^I} and $X^{\tau^{A/I}}$ denote the imprimitivity bimodules for inducing representations from $A \rtimes_{\alpha,\tau} H$, $I \rtimes_{\alpha,\tau} H$ and $(A/I) \rtimes_{\alpha,\tau} H$, respectively. Let $Y^\tau = X^\tau(I \rtimes_{\alpha,\tau} G)$. Then Y^τ is naturally isomorphic to X^{τ^I} and X^τ/Y^τ is naturally isomorphic to $X^{\tau^{A/I}}$.*

Since the lifted twisted actions of Example 1.1.3 play a very important role later in this work, it is useful to state the following proposition which follows directly from the definitions.

PROPOSITION 1.4.5. *Suppose that β is an action of $\tilde{G} = G/N$ on B and let $(\beta \circ q, 1_N)$ denote the twisted action of G on B which is lifted from β. Let H be a closed subgroup of G containing N and let X^{1_N} denote the imprimitivity bimodule for inducing representations from $B \rtimes_{\beta \circ q, 1_N} H$ to $B \rtimes_{\beta, 1_N} G$. Then*

X^{1_N} is identical to the imprimitivity bimodule Y for inducing representations from $B \rtimes_\beta \tilde{H}$ to $B \rtimes_\beta \tilde{G}$, if we identify $B \rtimes_\beta \tilde{H}$ with $B \rtimes_{\beta \circ q, 1_N} H$ and $B \rtimes_\beta \tilde{G}$ with $B \rtimes_{\beta \circ q, 1_N} G$ in the canonical way.

We are now going to give some relations between inducing and restricting representations of twisted crossed products. For this recall that if (A, G, α, τ) is a twisted covariant system and $H \supseteq N_\tau$ is a closed subgroup of G, then the restriction $\mathrm{res}_H^G(\pi \times U)$ of a representation $\pi \times U \in \mathrm{Rep}(A \rtimes_{\alpha,\tau} G)$ to $A \rtimes_{\alpha,\tau} H$ is defined by

$$\mathrm{res}_H^G(\pi \times U) = \pi \times U|_H,$$

where $U|_H$ denotes the restriction of U to H. Note that $\mathrm{res}_H^G(\pi \times U)$ can also be defined by first extending $\pi \times U$ to $\mathcal{M}(A \rtimes_{\alpha,\tau} G)$, and then embedding $A \rtimes_{\alpha,\tau} H$ into $\mathcal{M}(A \rtimes_{\alpha,\tau} G)$ via the integrated form of the canonical embeddings i_A and $i_H = i_G|_H$ of A and H in $A \rtimes_{\alpha,\tau} G$. Since extension of representations to multiplier algebras is continuous we obtain the following proposition (see [33, Section 3]).

PROPOSITION 1.4.6. *Let (A, G, α, τ) and H be as above. Then the maps*

$$\mathrm{ind}_H^G : \mathrm{Rep}(A \rtimes_{\alpha,\tau} H) \to \mathrm{Rep}(A \rtimes_{\alpha,\tau} G); \rho \times V \mapsto \mathrm{ind}_H^G(\rho \times V)$$

and

$$\mathrm{res}_H^G : \mathrm{Rep}(A \rtimes_{\alpha,\tau} G) \to \mathrm{Rep}(A \rtimes_{\alpha,\tau} H); \pi \times U \mapsto \mathrm{res}_H^G(\pi \times U)$$

are continuous.

We will see later that a similar result holds for induction and restriction from and to varying subgroups of G.

Before we state the next result recall that if $\rho \times V$ is a representation of $A \rtimes_{\alpha,\tau} H$ and $s \in G$, then we may define a representation $^s(\rho \times V)$ of $A \rtimes_{\alpha,\tau} H^s$, $H^s = sHs^{-1}$, by

$$^s(\rho \times V) = \rho \circ \alpha_{s^{-1}} \times V^s,$$

where the unitary representation V^s of H^s is defined by $V_h^s = V_{s^{-1}hs}$ for all $h \in H^s$. Note that in case where H is normal in G, $^s(\rho \times V)$ is just $(\rho \times V) \circ \gamma_{s^{-1}}^H$, where γ^H denotes the decomposition action of G on $A \rtimes_{\alpha,\tau} H$ as defined in Example 1.1.1. The following proposition shows that inducing representations of twisted crossed products is equivariant with respect to this G-action. For the proof see [33, Lemma 10] in case where H and L are normal in G, or [11, Proposition 1.12] for H and L arbitrary.

PROPOSITION 1.4.7. *Let (A, G, α, τ) be a twisted covariant system and let $L \supseteq H \supseteq N_\tau$ be closed subgroups of G. Then $^s(\mathrm{ind}_H^L(\rho \times V))$ is unitarily equivalent to $\mathrm{ind}_{H^s}^G \, {}^s(\rho \times V)$ for all $\rho \times V \in \mathrm{Rep}(A \rtimes_{\alpha,\tau} H)$ and $s \in G$.*

For the proof of the next result see for instance [11, Corollary 1.11].

PROPOSITION 1.4.8. *Suppose that* $H \supseteq N \supseteq N_\tau$ *are closed subgroups of* G *such that* N *is normal in* G *and let* $\rho \times V$ *be a covariant representation of* $A \rtimes_{\alpha,\tau} H$. *Then*

$$\ker(\mathrm{res}_N^G(\mathrm{ind}_H^G(\rho \times V))) = \bigcap_{s \in G} \ker(\,^s(\mathrm{res}_N^H(\rho \times V))).$$

Recall now that for any covariant representation (π, U) of (A, G, α, τ) on a Hilbert space $\mathcal{H}_\pi$ and for any unitary representation W of $\tilde{G} = G/N_\tau$ on $\mathcal{H}_W$ we may form the *tensor product*

$$W \otimes (\pi, U) = (1 \otimes \pi, W \otimes U),$$

which is a covariant representation of (A, G, α, τ) on $\mathcal{H}_W \otimes \mathcal{H}_\pi$. We will usually denote its integrated form by $W \otimes (\pi \times U)$. Note that it follows from similar arguments as used by Fell in [23] that

$$\mathrm{Rep}(C^*(\tilde{G})) \times \mathrm{Rep}(A \rtimes_{\alpha,\tau} G) \to \mathrm{Rep}(A \rtimes_{\alpha,\tau} G); (W, \pi \times U) \mapsto W \otimes (\pi \times U)$$

preserves weak containment in both variables (see [11, Proposition 1.2]). The following proposition is [11, Proposition 1.4].

PROPOSITION 1.4.9. *Let* (A, G, α, τ) *be a twisted covariant system and* H *a closed subgroup of* G.
  (1) *If* (ρ, V) *is a covariant representation of* (A, H, α, τ) *and* U *is a unitary representation of* $\tilde{G}$, *then* $U \otimes (\mathrm{ind}_H^G(\rho \times V))$ *is unitarily equivalent to* $\mathrm{ind}_H^G(U|_H \otimes (\rho \times V))$.
  (2) *If* V *is a unitary representation of* $\tilde{H} = H/N_\tau$ *and* (π, U) *is a covariant representation of* (A, G, α, τ), *then* $(\mathrm{ind}_H^G V) \otimes (\pi \times U)$ *is unitarily equivalent to* $\mathrm{ind}_H^G(V \otimes (\pi \times U|_H))$.

In particular, if (π, U) *is any covariant representation of* (A, G, α, τ) *and* N *is a closed normal subgroup of* G *containing* N_τ, *then* $\mathrm{ind}_N^G(\pi \times U|_N)$ *is unitarily equivalent to* $(\mathrm{ind}_N^G 1_N) \otimes (\pi \times U) = \lambda_{G/N} \otimes (\pi \times U)$, *where* $\lambda_{G/N}$ *denotes the left regular representation of* G/N.

If A is a C^*-algebra, then by the Dauns-Hoffmann theorem (see [50, Corollary 4.4.8]) there exists for each element $\varphi \in C^b(\widehat{A})$, the space of bounded continuous functions on $\widehat{A}$, a unique element z_φ in the center $\mathcal{Z}(A)$ of $\mathcal{M}(A)$ such that $\rho(z_\varphi) = \varphi(\rho)\,\mathrm{Id}_{\mathcal{H}_\rho}$ for all $\rho \in \widehat{A}$, and the map $\varphi \mapsto z_\varphi$ is an isomorphism from $C^b(\hat{A})$ onto $\mathcal{Z}(A)$.

Suppose now that (A, G, α) is a covariant system and that Ω is a locally compact G-space such that there exists a continuous G-equivariant map $R : \hat{A} \to \Omega$. Then we define a map

$$\Psi_R : C_0(\Omega) \odot A \to A; \Psi_R(\varphi \otimes a) = (z_{\varphi \circ R})a.$$

Then we have the following lemma (see [12, Lemma 6]).

LEMMA 1.4.10. Ψ_R extends to a $*$-homomorphism from $C_0(\Omega, A)$ onto A which is G-equivariant with respect to the diagonal action of G on $C_0(\Omega, A)$. Moreover, if $\rho \in \widehat{A}$, then $\Psi^*(\rho) = \rho \circ \Psi_R$ is equal to the pair $(R(\rho), \rho) \in \Omega \times \widehat{A} = C_0(\Omega, A)\widehat{\ }$.

Suppose now that (A, G, α, τ) is a twisted covariant system and that H is a closed subgroup of G such that there exists a continuous G-equivariant map R from $\mathrm{Prim}(A)$ onto G/H. Then it is easily seen that Ψ_R transports $\tilde{\tau}$ onto τ. Thus, Ψ_R implements a canonical surjective $*$-homomorphism, say Ψ, from $B^\tau = C_0(G/H, A) \rtimes_{\tilde{\alpha}, \tilde{\tau}} G$ onto $A \rtimes_{\alpha, \tau} G$. The following theorem of Green's [33, Theorem 17] is one of the most fundamental results in the representation theory of twisted crossed products.

THEOREM 1.4.11 (GREEN). Suppose that (A, G, α, τ), H, R and Ψ are as above. Let $I = \ker R^{-1}(\{\dot{e}\})$, $J = I \rtimes_{\alpha, \tau} H \subseteq C^\tau$ and $K = \ker \Psi \subseteq B^\tau$. Then K is the ideal induced from J via X^τ. Thus $X^\tau/X^\tau J$ becomes a $(A/I) \rtimes_{\alpha, \tau} H - A \rtimes_{\alpha, \tau} G$ imprimitivity bimodule. Moreover, if $\rho \times V \in \mathrm{Rep}((A/I) \rtimes_{\alpha, \tau} H)$, then Ψ carries the representation $\mathrm{ind}^{X^\tau}(\rho \times V)$ onto $\mathrm{ind}_H^G(\rho \times V)$, viewing $\rho \times V$ as a representation of $A \rtimes_{\alpha, \tau} H$ in the canonical way. Thus, induction from H to G establishes a homeomorphism between $((A/I) \rtimes_{\alpha, \tau} H)\widehat{\ }$ and $(A \rtimes_{\alpha, \tau} G)\widehat{\ }$.

Using decomposition of crossed products by G-invariant ideals as described in Example 1.1.2 the theorem above gives the so-called *first step of the Mackey machine* as stated below. A first version of this result has been proved by Mackey [38, 39] in order to find a procedure to describe the set of all equivalence classes of irreducible representations of a separable locally compact group. If F is a locally closed subset of $\mathrm{Prim}(A)$ (or $\widehat{A}$), then, as in Example 1.1.2, A_F denotes the subquotient of A corresponding to F.

THEOREM 1.4.12 (MACKEY-GREEN). Let (A, G, α, τ) be a twisted covariant system and let $P \in \mathrm{Prim}(A)$ such that the G-orbit $G(P) = \{\alpha_s(P); s \in G\}$ is locally closed. Assume further that the canonical map $G/S_P \to G(P); \dot{s} \mapsto \alpha_s(P)$ is a homeomorphism, where $S_P = \{s \in G; \alpha_s(P) = P\}$ denotes the stabilizer of P. Then $A_{G(P)} \rtimes_{\alpha, \tau} G$ is Morita equivalent to $A_P \rtimes_{\alpha, \tau} S_P$, where $A_P = A_{\{P\}}$ is the subquotient of A determined by the locally closed one-point set $\{P\} \subseteq \mathrm{Prim}(A)$. In particular, the map

$$\mathrm{ind}_{S_P}^G : (A_P \rtimes_{\alpha, \tau} S_P)\widehat{\ } \to (A_{G(P)} \rtimes_{\alpha, \tau} G)\widehat{\ }; \rho \times V \mapsto \mathrm{ind}_{S_P}^G(\rho \times V)$$

is a homeomorphism.

Note that the assumptions in the theorem above are satisfied if (A, G, α, τ) is separable, A is type I and $G(P)$ is locally closed in $\mathrm{Prim}(A)$ (see [63, Proposition 7.1]).

For any topological G-space M the *quasi-orbit* $\mathcal{Q}(m)$ of an element $m \in M$ consists of the set of all $m' \in M$ such that $\overline{G(m')} = \overline{G(m)}$, where $G(m) = \{sm; s \in G\}$ denotes the G-orbit of m. The quotient space $\mathcal{Q}_G(M)$ consisting of

all G-quasi orbits in M is called the *quasi-orbit space of M*. The proof of the
following proposition is a consequence of [**33**, lemma on p.221 and Corollary 19].

PROPOSITION 1.4.13. *Suppose that* (A, G, α, τ) *is a separable twisted covariant system. Then the map* $P \to \mathcal{Q}(P)$ *sending* $P \in \mathrm{Prim}(A)$ *to its quasi-orbit is open and the map*

$$\mathcal{Q}_G(\mathrm{Prim}(A)) \to \mathcal{I}(A); \mathcal{Q}(P) \mapsto \ker \mathcal{Q}(P) = \bigcap_{s \in G} \alpha_s(P)$$

is a homeomorphism onto its image. Moreover, for any $\pi \times U \in (A \rtimes_{\alpha,\tau} G)\hat{}$
there exists a unique quasi-orbit $\mathrm{Res}^G(\pi \times U) \in \mathcal{Q}_G(\mathrm{Prim}(A))$ *such that* $\ker \pi = \ker \mathrm{Res}^G(\pi \times U)$ *and the resulting map* $\mathrm{Res}^G : (A \rtimes_{\alpha,\tau} G)\hat{} \to \mathcal{Q}_G(\mathrm{Prim}(A))$ *is continuous.*

If α is an action of G on A, then the quasi-orbit spaces $\mathcal{Q}_G(\mathrm{Prim}(A))$ and
$\mathcal{Q}_G(\hat{A})$ coincide. This follows directly from the definition of the topologies on
$\hat{A}$ and $\mathrm{Prim}(A)$. It is a consequence of Proposition 1.4.13 and Theorem 1.4.12
that, if (A, G, α, τ) is a separable twisted system and all G-orbits in $\mathrm{Prim}(A)$
are locally closed, then every primitive ideal of $A \rtimes_{\alpha,\tau} G$ can be obtained as
the kernel of a representation induced from a stabilizer of some $P \in \mathrm{Prim}(A)$.
This result has been generalized substantially by Gootman and Rosenberg [**31**],
extending and using earlier work of Sauvageot [**67**]. Note that most results in
our later chapters depend heavily on this result.

THEOREM 1.4.14 (GOOTMAN-ROSENBERG). *Suppose that* (A, G, α, τ) *is a separable covariant system such that* $\tilde{G} = G/N_\tau$ *is amenable. If* $P \in \mathrm{Prim}(A)$
and $\rho \times V \in (A \rtimes_{\alpha,\tau} S_P)\hat{}$ *such that* $\ker \rho = P$, *then* $\ker(\mathrm{ind}_{S_P}^G(\rho \times V))$ *is a primitive ideal of* $A \rtimes_{\alpha,\tau} G$ *and every primitive ideal can be obtained in this way. Moreover, if* $\rho \times V$ *is as above, and if* $\pi \times U \in (A \rtimes_{\alpha,\tau} G)\hat{}$ *such that* $\ker(\pi \times U) = \ker(\mathrm{ind}_{S_P}^G(\rho \times V))$, *then* $\mathrm{Res}^G(\pi \times U) = \mathcal{Q}(P)$.

The last two theorems show that in many situations the problem of describing
the set of all primitive ideals of a twisted crossed product reduces to the problem
of describing the primitive ideals of the twisted crossed products $A_P \rtimes_{\alpha,\tau} S_P$,
provided $\{P\}$ is locally closed for all $P \in \mathrm{Prim}(A)$. At least in case where A
is type I, there is a procedure which enables us to describe all primitive ideals of these subsystems. This procedure is usually called the *second step of the
Mackey-Green machine* or just *Mackey's little group method*, and we are now going to recall the basic ingrediences of this procedure in the situation of separable
systems (A, G, α, τ) (for a more general version see [**33**, Theorem 18]).

If A is type I, then $\hat{A} \cong \mathrm{Prim}\, A$ and all points in $\hat{A}$ are locally closed. Moreover, for each $\rho \in \hat{A}$ the subquotient A_ρ of A defined by $\{\rho\}$ is canonically
isomorphic to the compact operators $\mathcal{K}(\mathcal{H}_\rho)$ on $\mathcal{H}_\rho$. Thus the investigation of
$A_\rho \rtimes_{\alpha,\tau} S_\rho$ reduces to the investigation of twisted covariant systems $\mathcal{K} \rtimes_{\alpha,\tau} G$,
where $\mathcal{K}$ denotes the compact operators on some (possibly finite dimensional)

separable Hilbert space $\mathcal{H}$. To this end recall that $\mathrm{Aut}(\mathcal{K})$ is canonically homeomorphic to the projective unitary group $\mathcal{P}\mathcal{U} = \mathcal{U}/\mathbb{T}1$ via conjugation, where $\mathcal{U} = \mathcal{U}(\mathcal{H})$ denotes the group of unitary operators on $\mathcal{H}$ equipped with the strong operator topology. We define

$$G' = \{(s,u) \in G \times \mathcal{U}; \alpha_s = \mathrm{Ad}\, u\}.$$

If we identify $\mathbb{T}$ with the subgroup $(e, \mathbb{T}1)$ of G' we see that G' is a central extension of G by $\mathbb{T}$.

Now let $c : G \to G'$ be a Borel section with $c(n) = (n, \tau_n)$ for all $n \in N_\tau$. Then

$$\omega(s,t) = c(s)c(t)c(st)^{-1}, \quad s,t \in G$$

is contained in the center $\mathbb{T}$ of G'. Thus ω may be viewed as a Borel map of $G \times G$ into $\mathbb{T}$. In fact, easy calculations show that ω is a cocycle in $Z^2(G, \mathbb{T})$ (compare with Example 1.1.4). If G^ω denotes the central extension of G by $\mathbb{T}$ defined by ω, then G^ω is isomorphic to G' via the map $(s, z) \mapsto c(s)z$. Note that by the construction of c, $\omega(sn, tm) = 1$ for all $n, m \in N_\tau$. Thus ω is lifted from a multiplier $\tilde{\omega}$ of $Z^2(\tilde{G}, \mathbb{T})$, $\tilde{G} = G/N_\tau$. Note also that ω, and hence also $\tilde{\omega}$, is trivial if and only if the section $c : G \to G'$ can be chosen as a continuous homomorphism, which in turn is true if and only if we find a unitary representation U of G on $\mathcal{H}$ such that

$$\alpha_s = \mathrm{Ad}\, U_s \quad \text{and} \quad U_n = \tau_n$$

for all $n \in N_\tau$ and $s \in G$, i.e. (id, U) is a covariant representation of $(\mathcal{K}, G, \alpha, \tau)$ (put $U = p \circ c$, where $p : G' \to \mathcal{U}$ denotes the projection on the second factor, or, in the other direction, take $c(s) = (s, U_s)$.) Thus it makes sense to call $\tilde{\omega}$ the *Mackey obstruction for extending* id *to a covariant representation of* $(\mathcal{K}, G, \alpha, \tau)$, or just the *Mackey obstruction* of $(\mathcal{K}, G, \alpha, \tau)$.

In any case, if we define $L_s = p(c(s))$, then $L : G \to \mathcal{U}$ is an ω-representation of G such that

(1) $L_n = \tau_n$ for all $n \in N_\tau$ and
(2) $\alpha_s = \mathrm{Ad}\, L_s$ for all $s \in G$.

We say that (id, L) is an ω-*covariant representation* of $(\mathcal{K}, G, \alpha, \tau)$.

The following result is the second step of the Mackey machine. Although the result is well known in much more generality (see [**33**, Theorem 18]) we will present here a very elementary proof, which, in our opinion, gives a clearer picture of the theory than the original proofs.

THEOREM 1.4.15 (MACKEY-TAKESAKI-GREEN). *Let $[\tilde{\omega}] \in H^2(\tilde{G}, \mathbb{T})$ be the Mackey obstruction of the separable twisted system $(\mathcal{K}, G, \alpha, \tau)$ and let $L : G \to \mathcal{U}(\mathcal{H})$ be an ω-representation of G which extends the identity $\mathrm{id} : \mathcal{K} \to \mathcal{K}$ to the ω-covariant representation (id, L) of $(\mathcal{K}, G, \alpha, \tau)$. Then the following is true:*

(1) $\mathcal{K} \rtimes_{\alpha, \tau} G$ *is naturally isomorphic to* $C^*(\tilde{G}, \tilde{\omega}^{-1}) \otimes \mathcal{K}$, *where $\tilde{\omega}^{-1} \in Z^2(\tilde{G}, \mathbb{T})$ denotes the inverse multiplier of $\tilde{\omega}$.*

(2) *The corresponding homeomorphism between* $C^*(\tilde{G}, \tilde{\omega}^{-1})\hat{}$ *and* $(\mathcal{K} \rtimes_{\alpha, \tau} G)\hat{}$ *is given by* $R \mapsto R \otimes (\mathrm{id} \times L)$, *where* R *runs through all equivalence classes of irreducible* $\tilde{\omega}^{-1}$ *representations of* $\tilde{G}$.

PROOF. Let $\tau'' : \mathbb{T} \to \mathbb{T}$ denote the canonical twisting map for the trivial action of $G'' = G^{\tilde{\omega}^{-1}}$ on $\mathbb{C}$ such that $C^*(\tilde{G}, \tilde{\omega}^{-1}) = \mathbb{C} \rtimes_{\mathrm{id}, \tau''} G''$ (see Example 1.1.4). We define

$$\Phi : L^1(G'', \mathbb{C}, \tau'') \odot \mathcal{K} \to L^1(G, \mathcal{K}, \tau); \quad \Phi(f \otimes k)(s) = f(\tilde{s}, 1)kL_s^*,$$

$s \in G$, $\tilde{s} = sN_\tau$. Straightforward computations show that Φ is a $*$-isometry, with respect to the canonical 1-norm on $L^1(G'', \mathbb{C}, \tau'') \odot \mathcal{K}$, viewed as a subspace of $L^1(G'', \mathbb{C} \otimes \mathcal{K}, \tau'' \otimes 1)$. Thus it will follow that Φ extends to an isomorphism between $C^*(\tilde{G}, \tilde{\omega}^{-1}) \otimes \mathcal{K}$ and $\mathcal{K} \rtimes_{\alpha, \tau} G$, as soon as we show that $\Phi(L^1(G'', \mathbb{C}, \tau'') \odot \mathcal{K})$ is dense in $L^1(G, \mathcal{K}, \tau)$.

In order to see this let $G' = G^{\omega^{-1}}$ denote the central extension of G by $\mathbb{T}$ defined by ω^{-1}, and let $\tau' : \mathbb{T} \to \mathbb{T}$ denote the canonical twisting map for G'. Define $\Phi' : L^1(G', \mathbb{C}, \tau') \otimes \mathcal{K} \to L^1(G, \mathcal{K})$ by $\Phi'(f \otimes k)(s) = f(s, 1)kL_s^*$. It is trivially seen that Φ' has dense image in $L^1(G, \mathcal{K})$. Now let $\Psi^\omega : L^1(G', \mathbb{C}, \tau') \to L^1(G'', \mathbb{C}, \tau'')$ be defined by

$$(\Psi^\omega f)(\tilde{s}, z) = \int_{N_\tau} f(sn, z) \, dn$$

and let $\Psi : L^1(G, \mathcal{K}) \to L^1(G, \mathcal{K}, \tau)$ denote the canonical homomorphism. Easy calculations show that

$$\Phi(\Psi^\omega \otimes \mathrm{id}(f \otimes k)) = \Psi(\Phi'(f \otimes k))$$

for all $f \otimes k \in L^1(G', \mathbb{C}, \tau') \odot \mathcal{K}$. Thus $\Phi(L^1(G'', \mathbb{C}, \tau'') \odot \mathcal{K})$ is dense in $L^1(G, \mathcal{K}, \tau)$ since Φ' has dense image in $L^1(G, \mathcal{K})$.

Now let R' be a unitary representation of G'' such that R' restricted to $\mathbb{T}$ is a multiple of the identity character and let R denote the $\tilde{\omega}^{-1}$-representation of $\tilde{G}$ defined by $R_{\tilde{s}} = R'_{(\tilde{s}, 1)}$. Then we compute

$$(1_{\mathcal{H}_R} \otimes \mathrm{id}) \times (R \otimes L)(\Phi(f \otimes k)) = \int_{\tilde{G}} 1_{\mathcal{H}_R} \otimes \mathrm{id}(\Phi(f \otimes k)(s))R_{\tilde{s}} \otimes L_s \, d\tilde{s}$$

$$= \int_{\tilde{G}} f(\tilde{s}, 1)(R_{\tilde{s}} \otimes k) \, d\tilde{s} = \left(\int_{\tilde{G}} f(\tilde{s}, 1)R'_{(\tilde{s}, 1)} \, d\tilde{s} \right) \otimes k = R' \otimes \mathrm{id}(f \otimes k)$$

for all $f \otimes k \in L^1(G'', \mathbb{C}, \tau'') \odot \mathcal{K}$. This completes the proof. $\square$

Combining this result with Green's theorem and the Gootman-Rosenberg theorem we obtain

PROPOSITION 1.4.16. *Suppose that* (A, G, α, τ) *is a separable twisted covariant system such that* A *is type I. For each* $\rho \in \hat{A}$ *let* S_ρ *denote the stabilizer of* ρ, *and let* $\tilde{S}_\rho = S_\rho/N_\tau$. *Then there exists a multiplier* $\tilde{\omega}_\rho \in Z^2(\tilde{S}_\rho, \mathbb{T})$, *unique up to similarity, such that*

(1) If ω_ρ is the pull-back of $\tilde{\omega}_\rho$ to S_ρ, then there exists an ω_ρ-representation L of S_ω in $\mathcal{H}_\rho$ such that (ρ, L) becomes a ω_ρ-covariant representation of (A, G, α, τ).

(2) If $(\tilde{S}_\rho, \tilde{\omega}_\rho^{-1})\widehat{\ }$ denotes the set of equivalence classes of irreducible $\tilde{\omega}_\rho^{-1}$-representations of $\tilde{S}_\rho$, canonically identified with $C^*(\tilde{S}_\rho, \tilde{\omega}_\rho^{-1})\widehat{\ }$. Then $(\tilde{S}_\rho, \tilde{\omega}_\rho^{-1})\widehat{\ }$ is homeomorphic to $(A_\rho \rtimes_{\alpha, \tau} S_\rho)\widehat{\ }$ via the map $R \mapsto R \otimes (\rho \times L)$.

(3) If $G(\rho)$ is locally closed for all $\rho \in \widehat{A}$, or if $\tilde{G} = G/N_\tau$ is amenable, then every primitive ideal Q of $A \rtimes_{\alpha, \tau} G$ is the kernel of some induced representation $\mathrm{ind}_{S_\rho}^G (R \otimes (\rho \times L))$, for some $\rho \in \widehat{A}$ and some $R \in (\tilde{S}_\rho, \tilde{\omega}_\rho^{-1})\widehat{\ }$.

We close this section with

DEFINITION 1.4.17. Let (A, G, α, τ) be a separable twisted covariant system such that A is type I. For each $\rho \in \widehat{A}$ let $\tilde{\omega}_\rho \in Z^2(\tilde{S}_\rho, \mathbb{T})$ be as in Proposition 1.4.16. Then the class of $\tilde{\omega}_\rho$ in $H^2(\tilde{S}_\rho, \mathbb{T})$ is called the *Mackey obstruction* of (A, G, α, τ) at ρ, or, more simply, the *Mackey obstruction of ρ*.

CHAPTER 2

Morita equivalent twisted actions and duality

Although twisted actions can be handled almost as easy as ordinary actions, there are sometimes situations where the twisting map complicates things more than we like. This is especially true if we want to use results known for ordinary crossed products, but for which no twisted version is available in the literature. This problem was partly solved by Packer and Raeburn in [47] by showing that every twisted action is exterior equivalent to an ordinary action. This allows to translate many results known for ordinary systems to the twisted case. However, the notion of twisted actions used in [47] differs substantially from Green's notion we use in this work. Although it was shown that every twisted action in Green's sense may be transformed into a twisted action in the sense of Packer and Raeburn, this still does not solve all problems, since so far there is no analogue of the modern Mackey-Green machine for the twisted crossed products of [47], which makes it often difficult, if not impossible, to translate results which use the Mackey-Green machine.

In order to overcome this problem the author extended in [15] the notion of Morita equivalent actions as introduced by Combes in [6] to twisted actions and showed that every twisted action (α, τ) is Morita equivalent to an ordinary action of $\tilde{G} = G/N_\tau$. Moreover, it was also shown in [15] that Morita equivalence of twisted actions is fully compatible with the Mackey-Green machine, which solves the problem stated above. Apart from these advantages, it turns out that the notion of Morita equivalent actions is a very helpful tool in the investigations of this work. Thus, we are now going to recall the basic definitions and the main results as proved in [15]. We then proceed by recalling the main results of [16] concerning Morita equivalence and duality for abelian twisted systems.

2.1. Morita equivalent twisted actions

Suppose that A and B are C^*-algebras. If X is a $B - A$ imprimitivity bimodule, then $\mathrm{Aut}(X)$ denotes the set of Banach space automorphism u of X such that

$$u(\langle \xi, \eta \rangle_A \zeta) = \langle u(\xi), u(\eta) \rangle_A u(\zeta)$$

(or equivalently $u(\xi\langle\eta,\zeta\rangle_B) = u(\xi)\langle u(\eta), u(\zeta)\rangle_B)$ for all $\xi, \eta, \zeta \in X$. If G is a locally compact group, then an action of G on X is a strongly continuous homomorphism $u : G \to \mathrm{Aut}(X)$. Recall from [6] that two actions α and β of G on A and B, respectively, are called *Morita equivalent* if there exists an action u of G on X such that

$$\beta_s(\langle\xi,\eta\rangle_B) = \langle u_s(\xi), u_s(\eta)\rangle_B \ \text{ and } \ \alpha_s(\langle\xi,\eta\rangle_A) = \langle u_s(\xi), u_s(\eta)\rangle_A$$

for all $s \in G$ and $\xi, \eta \in X$. The pair (X, u) is then called a $\beta - \alpha$ *imprimitivity pair*. If (X, u) is a $\beta - \alpha$ imprimitivity pair, then the equivalence between $\mathrm{Rep}(A)$ and $\mathrm{Rep}(B)$ given by induction via X is easily seen to be G-equivariant. The definition of Morita equivalent actions extends to twisted actions in the following way (see [15, Section 1]):

DEFINITION 2.1.1. Suppose that (α, τ) and (β, σ) are twisted actions of G on the C^*-algebras A and B, respectively. Then (α, τ) and (β, σ) are called *Morita equivalent*, if $N_\tau = N_\sigma$ and there exists a $\beta - \alpha$ imprimitivity pair (X, u) such that

$$\sigma_n\xi = u_n(\xi)\tau_n$$

for all $\xi \in X$ and $n \in N_\tau$. The pair (X, u) is then called a $(\beta, \sigma) - (\alpha, \tau)$ *imprimitivity pair*. Moreover, if (α, τ) is a twisted action of G on A and β is an ordinary action of $\tilde{G} = G/N_\tau$ on B, then (α, τ) is called to be *Morita equivalent to β* if (α, τ) is Morita equivalent to the lifted twisted action $(\beta \circ q, 1_{N_\tau})$ of G on B.

Morita equivalence of twisted actions is an equivalence relation. If (α, τ) and (β, σ) are twisted actions on the separable stable C^*-algebras A and B, respectively, then it was shown in [15, Proposition 4] that (α, τ) and (β, σ) are Morita equivalent if and only if they are outer conjugate (see [15] for the definitions). If (α, τ) is a twisted action of G on A, and if $\mathcal{K}$ is the algebra of compact operators on a Hilbert space $\mathcal{H}$, then (α, τ) is Morita equivalent to the action $(\alpha \otimes \mathrm{id}, \tau \otimes 1)$ of G on $A \otimes \mathcal{K}$. In fact, the pair $(A \otimes \mathcal{H}, \alpha \otimes 1)$ is easily seen to be a $(\alpha \otimes \mathrm{id}, \tau \otimes 1) - (\alpha, \tau)$ imprimitivity pair.

The next proposition shows that Morita equivalence of twisted actions is inherited to actions on quotients by G-invariant ideals (see the remarks following [15, Definition 1]):

PROPOSITION 2.1.2. *Suppose that (X, u) is a $(\beta, \sigma) - (\alpha, \tau)$ imprimitivity pair, and let I be a closed G-invariant ideal of A. Then $Y = XI$ is a G-invariant submodule of X and (Y, u^Y) and $(X/Y, u^{X/Y})$ become $(\beta^{I^X}, \sigma^{I^X}) - (\alpha^I, \tau^I)$ and $(\beta^{B/I^X}, \sigma^{B/I^X}) - (\alpha^{A/I}, \tau^{A/I})$ imprimitivity pairs, respectively. Here u^Y and $u^{X/Y}$ denote the canonical actions of G on Y and X/Y, respectively.*

A twisted action (α, τ) of G on A is called *unitary* if there exists a strictly continuous homomorphism $v : G \to \mathcal{U}(A)$ such that $\alpha_s = \mathrm{Ad}\, v_s$ for all $s \in G$ and $\tau_n = v_n$ for all $n \in N_\tau$. Note that in case where (α, τ) is unitary via v, $A \rtimes_{\alpha,\tau} G$

is isomorphic to $C^*(\tilde{G}) \otimes_{max} A$, $\tilde{G} = G/N_\tau$, via the isomorphism given on the dense subalgebra $C_c(G, A, \tau)$ by

$$\Phi : C_c(G, A, \tau) \to C_c(\tilde{G}, A); \Phi(f)(\tilde{s}) = f(s)v_s,$$

where we identify $C^*(\tilde{G}) \otimes_{max} A$ with the crossed product $A \rtimes_{\mathrm{id}} \tilde{G}$ by the trivial action id. If (α, τ) is unitary, then defining an action u of G on the $A - A$ imprimitivity bimodule A by $u_s(a) = v_s a$ shows that (α, τ) is Morita equivalent to the trivial action of $\tilde{G} = G/N_\tau$ on A. The following proposition is [15, Proposition 3]

PROPOSITION 2.1.3. *Suppose that (X, u) is a $(\beta, \sigma) - (\alpha, \tau)$ imprimitivity pair. If (α, τ) is unitary via the strictly continuous homomorphism $v : G \to \mathcal{U}(A)$, then (β, σ) is unitary via the strictly continuous homomorphism $w : G \to \mathcal{U}(B)$ defined by the equation*

$$w_s \xi = u_s(\xi) v_s$$

for all $\xi \in X$ and $s \in G$. Thus, (α, τ) is unitary if and only if (β, σ) is unitary. In particular, (α, τ) is unitary if and only if (α, τ) is Morita equivalent to the trivial action of $\tilde{G}$ on A.

If (X, u) is a $\beta - \alpha$ imprimitivity pair, then the crossed products $B \rtimes_\beta G$ and $A \rtimes_\alpha G$ are Morita equivalent, too. This follows easily seen from setting $\mathcal{B} = C_c(G, B)$, $\mathcal{X} = C_c(G, X)$ and $\mathcal{A} = C_c(G, A)$ and defining $\mathcal{B}$- and $\mathcal{A}$-valued inner products on $\mathcal{X}$ and left and right actions of $\mathcal{B}$ and $\mathcal{A}$ on $\mathcal{X}$ by

$$f \cdot \xi(s) = \int_G f(t) u_s(\xi(t^{-1}s)) \, dt$$

$$\xi \cdot g(s) = \int_G \xi(t) \alpha_t(g(t^{-1}s)) \, dt$$

$$\langle \xi, \eta \rangle_{\mathcal{A}}(s) = \int_G \alpha_{t^{-1}}(\langle \xi(t), \eta(ts) \rangle_A) \, dt$$

$$\langle \xi, \eta \rangle_{\mathcal{B}}(s) = \int_G \Delta_G(s^{-1}t) \langle \xi(t), u_t(\eta(s^{-1}t)) \rangle_B) \, dt$$

for $f \in \mathcal{B}$, $g \in \mathcal{A}$ and $\xi, \eta \in \mathcal{X}$. The completion, say $X \rtimes_u G$, of $\mathcal{X}$ becomes a $B \rtimes_\beta G - A \rtimes_\alpha G$ imprimitivity bimodule (see [6]).

Apart from inducing representations $\pi \times U$ of $A \rtimes_\alpha G$ to $B \rtimes_\beta G$ via the imprimitivity bimodule $X \rtimes_u G$, there is an alternative procedure for inducing a covariant representation (π, U) directly to a covariant representation, say $\mathrm{ind}^{(X,u)}(\pi, U)$, of (B, G, β) by defining

$$\mathrm{ind}^{(X,u)}(\pi, U) = (\mathrm{ind}^X \pi, u \otimes U),$$

where $\mathrm{ind}^X \pi$ denotes the representation of B induced from π via X, and where $u \otimes U$ acts on $X \otimes_A \mathcal{H}_\pi$ by

$$(u \otimes U)_s(\xi \otimes v) = u_s(\xi) \otimes U_s(v)$$

for all $\xi \in X, v \in \mathcal{H}_\pi$ and $s \in G$. $\mathrm{ind}^{(X,u)}(\pi, U)$ is called the *representation of* (B, G, β) *induced from* (π, U) *via* (X, u). It is easily seen that the map $\Phi : (X \rtimes_u G) \otimes_{A \rtimes_\alpha G} \mathcal{H}_\pi \to X \otimes_A \mathcal{H}_\pi$ defined by

$$\Phi(\xi \otimes v) = \int_G \xi(s) U_s(v)\, ds, \quad \xi \in C_c(G, X), v \in \mathcal{H}_\pi,$$

intertwines the integrated form of $\mathrm{ind}^{(X,u)}(\pi, U)$ with $\mathrm{ind}^{X \rtimes_u G}(\pi \times U)$. Thus induction via $X \rtimes_u G$ is equivalent to induction via (X, u).

Suppose now that σ and τ are twisting maps for β and α, respectively, such that (X, u) is a $(\beta, \sigma) - (\alpha, \tau)$ imprimitivity system. Then it turns out that $\mathrm{ind}^{(X,u)}(\pi, U)$ preserves σ if and only if (π, U) preserves τ. Thus $X \rtimes_u G/(X \rtimes_u G)I_\tau$ may be completed to a $B \rtimes_{\beta,\sigma} G - A \rtimes_{\alpha,\tau} G$ imprimitivity bimodule, say $X \rtimes_{u,\tau} G$.

It is a direct consequence of the definitions that Morita equivalence is inherited by the restriction of actions to closed subgroups of G. The following result shows that the equivalence of the covariant representations of Morita equivalent systems is compatible with the procedure of inducing representations from subgroups. For the proof see [**15**, Theorem 3].

PROPOSITION 2.1.4. *Suppose that* (X, u) *is an* $(\alpha, \tau) - (\beta, \sigma)$ *imprimitivity pair, and let H be a closed subgroup of G containing N_τ. Then the diagram*

$$
\begin{array}{ccc}
\mathrm{Rep}(A \rtimes_{\alpha,\tau} H) & \xrightarrow{\;\mathrm{ind}^{(X, u|_H)}\;} & \mathrm{Rep}(B \rtimes_{\beta,\sigma} H) \\
{\scriptstyle \mathrm{ind}_H^G} \downarrow & & \downarrow {\scriptstyle \mathrm{ind}_H^G} \\
\mathrm{Rep}(A \rtimes_{\alpha,\tau} G) & \xrightarrow[\;\mathrm{ind}^{(X,u)}\;]{} & \mathrm{Rep}(B \rtimes_{\beta,\sigma} G)
\end{array}
$$

commutes.

Suppose now that (A, G, α, τ) is a separable twisted system such that A is type I. For $\rho \in \hat{A}$ let ω_ρ be a multiplier of the stabilizer S_ρ such that ρ can be extended to an ω_ρ-covariant representation (ρ, L) of $(A, S_\rho, \alpha, \tau)$. If (β, σ) is a twisted action of G on the C^*-algebra B which is Morita equivalent to (α, τ) via the imprimitivity pair (X, u), then it is easily verified that $(\mathrm{ind}^X \rho, u \otimes L)$ is a ω_ρ-covariant representation of (B, G, β, σ). Thus we obtain

PROPOSITION 2.1.5. *Let* (A, G, α, τ) *and* (B, G, β, σ) *be separable twisted covariant systems such that A and B are type I. If (α, τ) and (β, σ) are Morita equivalent, then the Mackey obstructions of the systems coincide. More precisely, if (X, u) is a $(\beta, \sigma) - (\alpha, \tau)$ imprimitivity pair, and if $\rho \in \hat{A}$, then the Mackey obstruction of ρ is equal to the Mackey obstruction of $\mathrm{ind}^X \rho$.*

Suppose now that (α, τ) is a twisted action of G on A and let M be a closed normal subgroup of G containing N_τ. As in the previous section let $B_0^\tau =$

$C_c(G \times G/M, A, \tau)$, $X_0^\tau = C_c(G, A, \tau)$ and $C_0^\tau = C_c(M, A, \tau)$. We define an action β of G/M on B^τ by

$$\beta_{\dot{s}}(F)(t, \dot{r}) = F(t, \dot{r}\dot{s}),$$

$s, t, r \in G$, $F \in B_0^\tau$. Furthermore, we define an action u of G on X^τ by

$$u_s(\xi)(t) = \Delta_{\tilde{G}}(s)\Delta_{G/M}(s)^{-1/2}\xi(ts),$$

$\xi \in X_0^\tau$. Then we obtain

PROPOSITION 2.1.6. *Assume that (A, G, α, τ) and M are as above and let (γ^M, τ^M) denote the decomposition twisted action of G on $C^\tau = A \rtimes_{\alpha, \tau} M$ and let β and u be the actions of G/M and G on B^τ and X^τ, respectively, as defined above. Then (X^τ, u) becomes a $\beta - (\gamma^M, \tau^M)$ imprimitivity pair.*

PROOF. The first part of the proof of [**15**, Theorem 1] shows that the proposition is true in case where τ is trivial. The general case follows from the easy verifiable fact that the canonical quotient maps $C \to C^\tau$, $X \to X^\tau$ and $B \to B^\tau$ transport the actions in the non-twisted case to the actions as defined above. $\square$

If $M = N_\tau$, then $C^\tau = A \rtimes_{\alpha, \tau} N_\tau$ is isomorphic to A via the map $C_c(N_\tau, A, \tau) \to A$; $f \mapsto f(e)$. This map transports the decomposition twisted action $(\gamma^{N_\tau}, \tau^{N_\tau})$ to (α, τ). Thus, as a corollary of the proposition above we get

THEOREM 2.1.7. *Let (α, τ) be a twisted action of G on A. Then (α, τ) is Morita equivalent to an ordinary action β of $\tilde{G}$ on a C^*-algebra B.*

Since the system $(B, \tilde{G}, \beta)$ has the same properties as the lifted system $(B, G, \beta \circ q, 1_{N_\tau})$ in all relevant respects, we can use Theorem 2.1.7 in many situations to replace a given twisted action of G by an ordinary action of $\tilde{G}$. *This will be our standard way to reduce problems to the non-twisted situation.* Before we close this section we have to prove one additional result which shows that decomposition actions for Morita equivalent systems are Morita equivalent, too.

PROPOSITION 2.1.8. *Assume that (X, u) is a $(\beta, \sigma) - (\alpha, \tau)$ imprimitivity pair and let L be a closed normal subgroup of G containing N_τ. Let $(\gamma^{L,\alpha}, \tau^{L,\alpha})$ and $(\gamma^{L,\beta}, \tau^{L,\beta})$ denote the decomposition twisted actions of G on $A \rtimes_{\alpha, \tau} L$ and $B \rtimes_{\beta, \sigma} L$, respectively. Then $(\gamma^{L,\alpha}, \tau^{L,\alpha})$ is Morita equivalent to $(\gamma^{L,\beta}, \tau^{L,\beta})$.*

PROOF. We assume first that τ and σ are trivial. We define an action v of G on the $B \rtimes_\beta L - A \rtimes_\alpha L$ imprimitivity bimodule $X \rtimes_u L$ by

$$v_s(\xi)(t) = \delta(s)u_s(\xi(s^{-1}ts))$$

for all $s \in G$ and $\xi \in C_c(L, X)$. We easily check

$$(v_s(\xi)\langle v_s(\eta), v_s(\zeta)\rangle_{A \rtimes_\alpha L})(t)$$

$$= \int_L \int_L v_s(\xi)(r)\alpha_{rl^{-1}}(\langle v_s(\eta)(l), v_s(\zeta)(lr^{-1}t)\rangle_A)\, d_L r\, d_L l$$

$$= \delta(s)^3 \int_L \int_L u_s(\xi(s^{-1}rs))\alpha_{rl^{-1}}(\langle u_s(\eta(s^{-1}ls)), u_s(\zeta(s^{-1}lr^{-1}ts))\rangle_A)\, d_L r\, d_L l$$

$$= \delta(s) \int_L \int_L u_s(\xi(r))\alpha_{srl^{-1}}(\langle \eta(l), \zeta(lr^{-1}s^{-1}ts)\rangle_A)\, d_L r\, d_L l$$

$$= \delta(s)u_s \left(\int_L \int_L \xi(r)\alpha_{rl^{-1}}(\langle \eta(l), \zeta(lr^{-1}s^{-1}ts)\rangle_A)\, d_L r\, d_L l \right)$$

$$= v_s(\xi\langle \eta, \zeta\rangle_{A \rtimes_\alpha L})(t)$$

for all $s \in G$ and $\xi, \eta, \zeta \in C_c(L, X)$. Similar but easier computations show that

$$\langle v_s(\xi), v_s(\eta)\rangle_{A \rtimes_\alpha L} = \gamma_s^{L,\alpha}(\langle \xi, \eta\rangle_{A \rtimes_\alpha L})$$

and

$$\langle v_s(\xi), v_s(\eta)\rangle_{B \rtimes_\beta L} = \gamma_s^{L,\beta}(\langle \xi, \eta\rangle_{B \rtimes_\beta L})$$

for all $s \in G$ and $\xi, \eta \in C_c(L, X)$. Thus it follows that each v_s extends to an automorphism of $X \rtimes_u L$, and that $(X \rtimes_u L, v)$ becomes a $\gamma^{L,\beta} - \gamma^{L,\alpha}$ imprimitivity pair.

Next we observe that the left action of $\tau_l^{L,\beta}$ and the right action of $\tau_l^{L,\alpha}$ on $\xi \in C_c(L, X)$ are given by

$$(\tau_l^{L,\beta}\xi)(s) = u_l(\xi(l^{-1}s)) \quad \text{and} \quad (\xi\tau_l^{L,\alpha})(s) = \Delta_L(l^{-1})\xi(sl^{-1})$$

for all $l \in L$ and $\xi \in C_c(L, X)$ (the first equation follows from computing the expression $(\tau_l^{L,\beta}(\langle \xi, \eta\rangle_{B \rtimes_\beta L}\zeta))(s)$ for $\xi, \eta, \zeta \in C_c(L, X)$, while the second follows directly from the definitions). Thus, since $\delta(l) = \Delta_L(l)$ for all $l \in L$, it follows that

$$\tau_l^{L,\beta}\xi = u_l(\xi)\tau_l^{L,\alpha}$$

for all $l \in L$ and $\xi \in C_c(L, X)$. Thus we see that $(X \rtimes_u L, v)$ is a $(\gamma^{L,\beta}, \tau^{L,\beta}) - (\gamma^{L,\alpha}, \tau^{L,\alpha})$ imprimitivity pair. The general case for non-trivial τ and σ follows now by passing over to the quotient $X \rtimes_{u,\tau} L$ of $X \rtimes_u L$. $\square$

2.2. Morita equivalence and duality for abelian twisted systems

If (A, G, α, τ) is a twisted covariant system such that $\tilde{G} = G/N_\tau$ is abelian, then we call (A, G, α, τ) an *abelian twisted system*, or just an *abelian system* if τ is trivial. Recall from [**33**, Section 7] that for any abelian twisted system (A, G, α, τ) there is a canonical action, denoted $\widehat{\alpha}$, of $\widehat{\tilde{G}}$ on the twisted crossed product $A \rtimes_{\alpha,\tau} G$ which is defined by

$$(\widehat{\alpha}_\chi f)(s) = \overline{\chi(s)}f(s)$$

for all $\chi \in \widehat{\widetilde{G}}$ and $f \in C_c(G, A, \tau)$. $\widehat{\alpha}$ is called the *dual action* of $\widehat{\widetilde{G}}$ on $A \rtimes_{\alpha,\tau} G$.

Suppose now that M is a closed subgroup of G containing N_τ. Let C^τ and B^τ denote the algebras $A \rtimes_{\alpha,\tau} M$ and $C_0(G/M, A) \rtimes_{\tilde{\alpha},\tilde{\tau}} G$, and let X^τ denote the canonical $B^\tau - C^\tau$ imprimitivity bimodule (see Section 1.4). Let $M^\perp$ denote the subgroup of $\widehat{\widetilde{G}}$ consisting of all characters on G which vanish on M, i.e. $M^\perp \cong \widehat{G/M}$. There is a canonical covariant representation (ϱ^M, W^M) of $(A \rtimes_{\alpha,\tau} G, M^\perp, \widehat{\alpha})$ into $\mathcal{M}(B^\tau)$ which is given by

$$(\varrho^M(f)F)(s, \dot{t}) = \int_{\tilde{G}} f(r)\alpha_r(F(r^{-1}s, r^{-1}t))\, d\tilde{r}$$

and

$$(W^M_\chi F)(s, \dot{t}) = \overline{\chi(t)}F(s, \dot{t})$$

for $F \in B^\tau_0$, $f \in C_c(G, A, \tau)$ and $\chi \in M^\perp$. Note that via the canonical identification of $\mathcal{M}(B^\tau)$ with the algebra $\mathcal{L}_{C^\tau}(X^\tau)$ of all bounded C^τ-linear operators on X^τ, (ϱ^M, W^M) may be identified with the covariant representation, also denoted (ϱ^M, W^M), of $(A \rtimes_{\alpha,\tau} G, M^\perp, \widehat{\alpha})$ on X^τ given by

$$\varrho^M(f)\xi = f * \xi \quad \text{and} \quad W^M_\chi \xi = \overline{\chi(s)}\xi(s)$$

for all $f \in C_c(G, A, \tau)$, $\xi \in X^\tau_0$ and $\chi \in M^\perp$. It was pointed out in [16], following some arguments used in [33, Section 7], that the integrated form $\varrho^M \times W^M$ defines an isomorphism from $(A \rtimes_{\alpha,\tau} G) \rtimes_{\widehat{\alpha}} M^\perp$ onto B^τ. Thus we see that X^τ becomes a $(A \rtimes_{\alpha,\tau} G) \rtimes_{\widehat{\alpha}} M^\perp - A \rtimes_{\alpha,\tau} M$ imprimitivity bimodule in a natural way.

There are canonical twisted actions of G and $\widehat{\widetilde{G}}$ on $(A \rtimes_{\alpha,\tau} G) \rtimes_{\widehat{\alpha}} M^\perp$ and $A \rtimes_{\alpha,\tau} M$, respectively, which are given as follows: The G-action on $A \rtimes_{\alpha,\tau} M$ is the decomposition action (γ^M, τ^M), and the twisted action of G on $(A \rtimes_{\alpha,\tau} G) \rtimes_{\widehat{\alpha}} M^\perp$ is the action $(\widehat{\widehat{\alpha}|_{M^\perp}} \circ q^M, 1_M)$ which is lifted from the double dual action $\widehat{\widehat{\alpha}|_{M^\perp}}$ of G/M on $(A \rtimes_{\alpha,\tau} G) \rtimes_{\widehat{\alpha}} M^\perp$. The twisted $\widehat{\widetilde{G}}$-actions on these crossed products are given by the decomposition action $(\gamma^{M^\perp}, \tau^{M^\perp})$ of $\widehat{\widetilde{G}}$ on $(A \rtimes_{\alpha,\tau} G) \rtimes_{\widehat{\alpha}} M^\perp$ and the twisted action $(\widehat{\alpha|_M} \circ q^{M^\perp}, 1_{M^\perp})$ on $A \rtimes_{\alpha,\tau} M$ which is lifted from the dual action $\widehat{\alpha|_M}$ of $\widehat{M}$ on $A \rtimes_{\alpha,\tau} M$.

The following proposition is [16, Proposition 3.4] together with [16, Lemma 3.6] (compare also with Proposition 2.1.6). It gives a very general version of duality for abelian twisted systems.

PROPOSITION 2.2.1. *Let* (A, G, α, τ), M *and* X^τ *be as above. Let* u *be the action of* G *on* X^τ *defined by*

$$(u_s(\xi))(t) = \xi(ts)$$

for $s, t \in G$, $\xi \in X^\tau_0$, *and let* $\widehat{u}$ *be the action of* $\widehat{\widetilde{G}}$ *on* X^τ *defined by*

$$(\widehat{u}_\chi \xi)(s) = \overline{\chi(s)}\xi(s).$$

Then (X^τ, u) is a $(\widehat{\widehat{\alpha}|_{M^\perp}} \circ q^M, 1_M) - (\gamma^M, \tau^M)$ imprimitivity pair and $(X^\tau, \widehat{u})$ is a $(\widehat{\alpha|_M} \circ q^{M^\perp}, 1_{M^\perp}) - (\widehat{\alpha|_M} \circ q^{M^\perp}, 1_{M^\perp})$ imprimitivity pair. In particular, (γ^M, τ^M) is Morita equivalent to $\widehat{\widehat{\alpha}|_{M^\perp}}$ and $(\gamma^{M^\perp}, \tau^{M^\perp})$ is Morita equivalent to $\widehat{\alpha|_M}$.

In the special case where $M = N_\tau$ we obtain the result that (α, τ) is Morita equivalent to the double dual action $\widehat{\widehat{\alpha}}$ of $\tilde{G}$ on $(A \rtimes_{\alpha,\tau} G) \rtimes_{\widehat{\alpha}} \widehat{\tilde{G}}$. Note that these results may be viewed as very general versions of the Takai duality theorem for abelian covariant systems, which says in its strongest version that the double dual action $\widehat{\widehat{\alpha}}$ is stably exterior equivalent to α (see also [**15**, Proposition 4]).

If (A, G, α, τ) and M are as above, then we will denote by

$$\mathrm{IND}^M : \mathrm{Rep}(A \rtimes_{\alpha,\tau} M) \to \mathrm{Rep}((A \rtimes_{\alpha,\tau} G) \rtimes_{\widehat{\alpha}} M^\perp)$$

the equivalence of the representation spaces given by induction via the imprimitivity bimodule X^τ. The following duality theorem for induction and restriction of representations of twisted abelian systems constitutes the main result of [**16**].

THEOREM 2.2.2. *Let (A, G, α, τ) be an abelian twisted system and let $N_\tau \subseteq M \subseteq H$ be closed subgroups of G. Then the diagrams*

$$
\begin{array}{ccc}
\mathrm{Rep}(A \rtimes_{\alpha,\tau} M) & \xrightarrow{\ \mathrm{IND}^M\ } & \mathrm{Rep}((A \rtimes_{\alpha,\tau} G) \rtimes_{\widehat{\alpha}} M^\perp) \\
{\scriptstyle \mathrm{ind}_M^H} \downarrow & & \downarrow {\scriptstyle \mathrm{res}_{H^\perp}^{M^\perp}} \\
\mathrm{Rep}(A \rtimes_{\alpha,\tau} H) & \xrightarrow{\ \mathrm{IND}^H\ } & \mathrm{Rep}((A \rtimes_{\alpha,\tau} G) \rtimes_{\widehat{\alpha}} H^\perp)
\end{array}
$$

and

$$
\begin{array}{ccc}
\mathrm{Rep}(A \rtimes_{\alpha,\tau} M) & \xrightarrow{\ \mathrm{IND}^M\ } & \mathrm{Rep}((A \rtimes_{\alpha,\tau} G) \rtimes_{\widehat{\alpha}} M^\perp) \\
{\scriptstyle \mathrm{res}_M^H} \uparrow & & \uparrow {\scriptstyle \mathrm{ind}_{H^\perp}^{M^\perp}} \\
\mathrm{Rep}(A \rtimes_{\alpha,\tau} H) & \xrightarrow{\ \mathrm{IND}^H\ } & \mathrm{Rep}((A \rtimes_{\alpha,\tau} G) \rtimes_{\widehat{\alpha}} H^\perp)
\end{array}
$$

commute.

CHAPTER 3

Representations of type I abelian twisted systems

In this chapter we are investigating in detail the structure of primitive ideal spaces and dual spaces of crossed products $A \rtimes_{\alpha, \tau} G$ of separable abelian twisted systems (A, G, α, τ), if A is assumed to be type I. To this end we will introduce the notion of maximally pointwise unitary subgroups for such systems. We will see that every primitive ideal of $A \rtimes_{\alpha, \tau} G$ is induced from such groups. We will also give a precise description whether $A \rtimes_{\alpha, \tau} G$ is type I in terms of the maximally pointwise unitary subgroups of the system, and give relations between the maximally pointwise unitary subgroups of a system (A, G, α, τ) and those of the dual system $(A \rtimes_{\alpha, \tau} G, \widehat{\widehat{G}}, \widehat{\alpha})$. We will also compute the Mackey obstructions of the dual system in terms of the Mackey obstructions of (A, G, α, τ). We should mention that the material of this chapter bases heavily on a result of Gootman [28, Theorem 2.3], which asserts that if an abelian groop G acts on a type I C^*-algebra A, then the Mackey obstructions are constant on the quasi-orbits in $\widehat{A}$. Some of the results of this chapter (in particular of Section 3.1) are well known – we tried to give the right references in all those situations – but we hope that the treatment given here will be helpful for the understanding of the following chapters.

3.1. Maximally pointwise unitary subgroups

Suppose that G is a separable abelian group and that $\omega \in Z^2(G, \mathbb{T})$ is a multiplier of G. Then ω defines a canonical homomorphism $h_\omega : G \to \widehat{G}$ by

$$h_\omega(s)(t) = \omega(s, t)\omega(t, s)^{-1}$$

for all $s, t \in G$. The kernel

$$\Sigma_\omega = \{s \in G; \omega(s, t) = \omega(t, s) \text{ for all } t \in G\}$$

of h_ω is called the *symmetry group* of ω. A multiplier ω is called *totally skew* if Σ_ω is trivial. Note that the image of h_ω is always a dense subgroup of $\widehat{G/\Sigma_\omega}$. The homomorphism h_ω and the symmetry group Σ_ω only depend on the class of ω in $H^2(G, \mathbb{T})$ and play a fundamental role in the representation theory of the twisted group algebra $C^*(G, \omega)$ defined by ω, and hence, via the second step of the Mackey machine, also for the representation theory of separable twisted abelian systems (A, G, α, τ) (see [**3**] for more details). For instance it is now a classical result of Baggett and Kleppner [**3**] that ω is type I, i.e., $C^*(G, \omega)$ is type I, if and only if $h_\omega(G)$ is all of $\widehat{G/\Sigma}$ (see also [**33**, Section 7]).

In this section we show that there is also another class of subgroups of G which has interesting properties with respect to the representation theory of (A, G, α, τ). We start by recalling from [**19**, Section 2] the definition of a maximally ω-trivial subgroup of G. Recall that a multiplier is called *trivial* if it is similar to the trivial multiplier $(s, t) \mapsto 1$.

DEFINITION 3.1.1. Let ω be a multiplier on the separable abelian group G. A closed subgroup H of G is called *maximally ω-trivial* if H is a maximal closed subgroup of G with respect to the property that the restriction $\omega_H = \omega|_{H \times H}$ is trivial.

The following lemma is well known (see for instance [**5**, Section 2], where maximally ω-trivial subgroups are called maximally isotropic). But for the readers convenience, we give the complete proof anyway.

LEMMA 3.1.2. *Let ω be a multiplier on the separable abelian group G. Then there exists a maximally ω-trivial subgroup H of G. Moreover, if H is maximally ω-trivial and if $h_\omega^H : G \to \widehat{H}$ denotes the composition of h_ω with the quotient map $\widehat{G} \to \widehat{H}$ given by restriction, then $\ker h_\omega^H = H$ and h_ω^H factors through a homomorphism from G/H onto a dense subgroup of $\widehat{H/\Sigma}_\omega$.*

PROOF. Let $\mathcal{M}$ denote the set of all closed subgroups L of G such that ω_L is trivial, and let $\mathcal{L}$ be a chain in $\mathcal{M}$. Then it follows easily from the continuity of h_ω that the restriction of ω to the closure K of $\cup_{L \in \mathcal{L}} L$ is trivial, too. Hence by Zorn's lemma there exists a maximal element H in $\mathcal{M}$, which just means that H is a maximally ω-trivial subgroup of G.

Suppose now that H is maximally ω-trivial and let h_ω^H be as in the lemma. Then it is clear that H is contained in the group kernel of h_ω^H. Assume now that there exists $s \notin H$ such that $h_\omega(s)(h) = 1$ for all $h \in H$. Then $h_\omega(s^n)(h) = (h_\omega(s)(h))^n = 1$ for all integers $n \in \mathbb{Z}$. Since $h_\omega(s^n)(s^m) = (h_\omega(s)(s))^{n+m} = 1$ for all $n, m \in \mathbb{Z}$ we now conclude from the continuity of h_ω in both variables that ω_L is trivial if L denotes the closed subgroup of G generated by H and s. But this contradicts the maximallity of H as an ω-trivial subgroup of G. Thus $H = \ker h_\omega^H$. The fact that $h_\omega^H(G)$ is a dense subgroup of $\widehat{H/\Sigma}_\omega$ follows easily from the fact that $h_\omega(G)$ is dense in $\widehat{G/\Sigma}_\omega$. $\square$

Note that maximally ω-trivial subgroups are not uniquely determined. In fact, two maximally ω-trivial subgroups H and H' of G need not even be isomorphic. As an example look at the multiplier ω on $\mathbb{Z} \times \mathbb{T}$ defined by $\omega((n, z), (m, w)) = z^m$, $n, m \in \mathbb{Z}$, $z, w \in \mathbb{T}$. Then it is easily seen that both, $\mathbb{Z}$ and $\mathbb{T}$, are maximally

ω-trivial subgroups on $\mathbb{Z} \times \mathbb{T}$. We are now going to define maximally ρ-unitary subgroups of G.

DEFINITION 3.1.3. Let (A, G, α, τ) be a twisted covariant system and let $\rho \in \widehat{A}$. A closed subgroup H of G is called *maximally ρ-unitary* if H is a maximal closed subgroup with respect to the property that there exists a unitary representation V of G on $\mathcal{H}_\rho$ such that (ρ, V) is a covariant representation of (A, H, α, τ).

PROPOSITION 3.1.4. *Suppose that (A, G, α, τ) is a separable abelian twisted system such that A is type I and let $\rho \in \widehat{A}$. The maximally ρ-unitary subgroups of G are exactly the pull-backs of the maximally $\tilde{\omega}_\rho$-trivial subgroups of $\tilde{S}_\rho = S/N_\tau$, where $\tilde{\omega}_\rho \in Z^2(\tilde{S}_\rho, \mathbb{T})$ is in the class of the Mackey obstruction of ρ. In particular, maximally ρ-unitary subgroups always exist.*

PROOF. Recall from the definition of the Mackey obstruction (see Section 1.4) that for any closed subgroup L of G there exists a unitary representation V of L such that (ρ, V) is a covariant representation of (A, L, α, τ) if and only if the Mackey obstruction of the system $(A_\rho, L, \alpha, \tau)$ is trivial, where, as usual, A_ρ denotes the subquotient of G defined by the locally closed one-point set $\{\rho\}$ in $\widehat{A}$. Since for any closed subgroup L of G the Mackey obstructions of (A, L, α, τ) are the restrictions of the Mackey obstructions of (A, G, α, τ) to the intersections of the stabilizers $\tilde{S}_\rho = S_\rho/N_\tau$ with $\tilde{L} = L/N_\tau$, it follows that H is a maximal ρ-unitary subgroup of G if and only if $\tilde{H} = H/N_\tau$ is a maximally $\tilde{\omega}_\rho$-trivial subgroup of $\tilde{S}_\rho$. Thus, the proof follows from Lemma 3.1.2. $\square$

Clearly, as for maximally ω-trivial subgroups there may be different and non-isomorphic maximally ρ-unitary subgroups of G. Before we give the next definition we have to recall a result due to Gootman which says that for separable abelian covariant systems (A, G, α) with A being type I the stability groups S_ρ and the Mackey obstructions $[\omega_\rho] \in H^2(S_\rho, \mathbb{T})$ are constant on quasi-orbits in $\widehat{A}$ (see [**28**, Lemma 2.2, Theorem 2.3]). Since the stability groups and the Mackey obstructions are invariant by passing over to Morita equivalent actions we conclude that the same result is true for separable abelian twisted systems. Thus, by Proposition 3.1.4, it is possible to choose the same maximally ρ-unitary subgroup H_ρ of G for all ρ in a fixed quasi-orbit in $\widehat{A}$.

DEFINITION 3.1.5. Let (A, G, α, τ) be a separable abelian twisted system such that A is type I and denote by $\mathfrak{K}(G)$ the set of all closed subgroups of G. A *choice of maximally pointwise unitary subgroups* for (A, G, α, τ) is a map $H : \widehat{A} \to \mathfrak{K}(G); \rho \mapsto H_\rho$ such that, for all $\rho \in \widehat{A}$, H_ρ is a maximally ρ-unitary subgroup of G and H is constant on quasi-orbits in $\widehat{A}$.

Remark 3.1.6. If (α, τ) is Morita equivalent to an action β of $\tilde{G} = G/N_\tau$ on B via an imprimitivity pair (X, u), then we know from Proposition 2.1.5 that the Mackey obstruction of $\mathrm{ind}^X \rho \in \widehat{B}$ is equal to the Mackey obstruction of ρ. Thus it follows that if $H : \widehat{A} \to \mathfrak{K}(G)$ is a choice of maximally pointwise unitary subgroups for (A, G, α, τ), then $\tilde{H} : \widehat{B} \to \mathcal{K}(G)$; $\tilde{H}_{\text{ind}\times \rho} = H_\rho/N_\tau$ is a choice of maximally pointwise unitary subgroups for $(B, \tilde{G}, \beta)$.

We are now going to describe the primitive ideal space of $A \rtimes_{\alpha,\tau} G$ with respect to a choice of maximally pointwise unitary subgroups.

THEOREM 3.1.7. *Suppose that* (A, G, α, τ) *is a separable abelian twisted system such that A is type I and let $H : \widehat{A} \to \mathcal{K}(G)$ be a choice of maximally pointwise unitary subgroups for* (A, G, α, τ). *Define*

$$\mathcal{R} = \{\rho \times U \in (A \rtimes_{\alpha,\tau} H_\rho)\widehat{} \, ; \rho \in \widehat{A}\}.$$

Then $\ker(\text{ind}^G_{H_\rho}(\rho \times U)) \in \text{Prim}(A \rtimes_{\alpha,\tau} G)$ *for all $\rho \times U \in \mathcal{R}$ and the resulting map*

$$\text{Ind} : \mathcal{R} \to \text{Prim}(A \rtimes_{\alpha,\tau} G); \rho \times U \mapsto \ker(\text{ind}^G_{H_\rho}(\rho \times U))$$

is surjective. Moreover, $\text{Ind}(\rho \times U) = \text{Ind}(\rho' \times U')$ *if and only if ρ and ρ' lie in the same quasi-orbit in $\widehat{A}$ and $\rho \times U$ and $\rho' \times U'$ lie in the same quasi-orbit in $(A \rtimes_{\alpha,\tau} H_\rho)\widehat{}$ under the canonical action γ^{H_ρ} of G on $(A \rtimes_{\alpha,\tau} H_\rho)\widehat{}$.*

Remark 3.1.8. Assume that (A, G, α, τ), H and $\mathcal{R}$ are as in Theorem 3.1.7. For each $\rho \in \widehat{A}$ let U^ρ denote a fixed unitary representation of H_ρ extending ρ to a covariant representation (ρ, U^ρ) of $(A, H_\rho, \alpha, \tau)$. Then we may write

$$\mathcal{R} = \{\rho \times \chi U^\rho; \rho \in \widehat{A}, \chi \in \widehat{\tilde{H}}_\rho\}.$$

For this observe that by the remarks preceding Proposition 2.1.3 each subsystem $A_\rho \rtimes_{\alpha,\tau} H_\rho$ is isomorphic to $C^*(\tilde{H}_\rho) \otimes A_\rho$ via the extension of the map

$$\Phi^\rho : C_c(H_\rho, A_\rho, \tau) \to C_c(\tilde{H}_\rho, A_\rho); \Phi^\rho(f)(\tilde{s}) = f(s)U_s$$

to $A_\rho \rtimes_{\alpha,\tau} H_\rho$. But it is easily seen that this isomorphism carries the pair $(\chi, \rho) \in \widehat{\tilde{H}}_\rho \times \{\rho\} = (C^*(\tilde{H}_\rho) \otimes A_\rho)\widehat{}$ onto $\rho \times \chi U^\rho \in (A_\rho \rtimes_{\alpha,\tau} H_\rho)\widehat{}$.

For the proof of Theorem 3.1.7 we need the following lemmas. The first states a commutation relation for ω-representations on abelian groups, which follows directly from the multiplier identities.

LEMMA 3.1.9. *Suppose that ω is a multiplier on the (separable) locally compact abelian group G, and let $L : G \to \mathcal{U}(\mathcal{H})$ be an ω-representation of G. Then*

$$L_s^* L_t = h_\omega(s)(t) L_t L_s^* \quad \text{for all } s, t \in G.$$

We use this commutation relation to prove the following lemma, which is fundamental for the proof of Theorem 3.1.7.

LEMMA 3.1.10. *Let (A, G, α, τ) be a separable abelian system such that A is type I and let $\rho \in \hat{A}$. Let $[\omega] \in H^2(\tilde{S}_\rho, \mathbb{T})$ denote the Mackey obstruction of ρ and let H be a maximally ρ-unitary subgroup of G. Moreover, let U be a unitary representation of H which extends ρ to a covariant representation of (A, H, α, τ). Then $\rho \times U \circ \gamma_s^H$ is unitarily equivalent to $\rho \times h_\omega^H(s)U$ for all $s \in S_\rho$, where γ^H denotes the decomposition action of G on $A \rtimes_{\alpha, \tau} H$ and $h_\omega^H(s) = h_\omega(\tilde{s})|_H$ for all $s \in S_\rho$. In particular, the stabilizer of $\rho \times U$ under the canonical action of G is equal to H.*

PROOF. By Proposition 2.1.8 we may pass over to a Morita equivalent action in order to assume that τ is trivial and $G = \tilde{G}$ is abelian. Since G is abelian we have

$$\rho \times U \circ \gamma_s^H(f) = \int_H \rho(\alpha_s(f(t)))U_t \, dt = \rho \circ \alpha_s \times U(f)$$

for all $f \in C_c(G, A)$ and $s \in G$. Thus the stabilizer of $\rho \times U$ must be contained in the stabilizer S_ρ of ρ.

Let $L : S_\rho \to \mathcal{U}(\mathcal{H}_\rho)$ be an ω-representation of S_ρ satisfying $\rho(\alpha_s(a)) = L_s\rho(a)L_s^*$. Since $\omega|_{H \times H}$ is trivial, we may assume by passing over to a similar multiplier that $L|_H$ is unitary, and, by multiplying L with an appropriate character of S_ρ, we may further assume that $L|_H = U$. Using Lemma 3.1.9 we obtain for all $f \in C_c(H, A)$ and $s \in S_\rho$:

$$(\rho \times U) \circ \gamma_s^H(f) = (\rho \times L|_H)(\gamma_s^H(f)) = \int_H \rho(\alpha_s(f(t)))L_t \, dt$$

$$= \int_H L_s\rho(f(t))L_s^*L_t \, dt = L_s \left(\int_H \rho(f(t))h_\omega(s)(t)U_t \, dt \right) L_s^*$$

$$= L_s(\rho \times h_\omega^H(s)U(f))L_s^*.$$

By Lemma 3.1.2 we know that the group kernel of h_ω^H is equal to H from which follows that for $s \in S_\rho$, $\rho \times U \circ \gamma_s^H$ is equivalent to $\rho \times U$ if and only if $s \in H$. This finishes the proof. $\square$

PROOF OF THEOREM 3.1.7. Since all assertions of the theorem are invariant under passing over to a Morita equivalent action, we may assume that τ is trivial. Let $(\rho \times U) \in \mathcal{R}$ and let $(\gamma^{H_\rho}, \tau^{H_\rho})$ denote the decomposition twisted action for H_ρ. By Lemma 3.1.10 we know that that the stabilizer of $\rho \times \chi U^\rho$ under this action is equal to H_ρ. Thus, it follows from the Gootman-Rosenberg theorem (Theorem 1.4.14) applied to the system $(A \rtimes_\alpha H_\rho, G, \gamma^{H_\rho}, \tau^{H_\rho})$ that $\ker(\text{ind}_{H_\rho}^G(\rho \times U))$ is a primitive ideal of $A \rtimes_\alpha G$.

In order to see that all primitive ideals arise in this way, let us start with a primitive ideal P of $A \rtimes_\alpha G$. Then, again by the Gootman-Rosenberg theorem, we find some $\rho \in \hat{A}$, and an irreducible representation $\pi \times V$ of $A_\rho \rtimes_\alpha S_\rho$, such that $P = \ker(\text{ind}_{S_\rho}^G(\pi \times V))$. Thus, by the theorem of induction in stages and by the continuity of induction (see Propositions 1.4.2 and 1.4.6), it is enough to show that $\ker(\pi \times V) = \ker(\text{ind}_{H_\rho}^{S_\rho}(\rho \times U))$ for some $\rho \times U \in \mathcal{R}$. For this we apply Lemma 3.1.10 to the system (A_ρ, S_ρ, α) in order to see that the action of S_ρ on $(A_\rho \rtimes_\alpha H_\rho)\widehat{\ }$ induced from the decomposition action γ^{H_ρ} has constant stabilizer H_ρ. Thus, using the Gootman-Rosenberg theorem a third time, we see that every primitive ideal of $A_\rho \rtimes_\alpha S_\rho$ is the kernel of $\mathrm{ind}_{H_\rho}^{S_\rho}(\rho \times U)$ for some $\rho \times U \in (A_\rho \rtimes_\alpha H_\rho)\widehat{\ } \subseteq \mathcal{R}$.

Assume now that $\rho \times U$ and $\rho' \times U'$ are elements of $\mathcal{R}$ such that $\ker(\mathrm{ind}_{H_\rho}^G(\rho \times U)) = \ker(\mathrm{ind}_{H_{\rho'}}^G(\rho' \times U'))$. Then it follows from the continuity of $\mathrm{res}_{\{e\}}^G$ and Proposition 1.4.8 that

$$\bigcap_{s \in G} \ker(\rho \circ \alpha_s) = \ker(\mathrm{res}_{\{e\}}^G(\mathrm{ind}_{H_\rho}^G(\rho \times U)))$$

$$= \ker(\mathrm{res}_{\{e\}}^G(\mathrm{ind}_{H_{\rho'}}^G(\rho' \times U'))) = \bigcap_{s \in G} \ker(\rho' \circ \alpha_s),$$

which just means that ρ and ρ' lie in the same quasi-orbit. In particular, this implies that $H_\rho = H_{\rho'}$. Thus, applying the same argument as above to the system $(A \rtimes_\alpha H_\rho, G, \gamma^{H_\rho}, \tau^{H_\rho})$, we see that $\rho \times U$ and $\rho' \times U'$ lie in the same quasi-orbit in $(A \rtimes_\alpha H_\rho)\widehat{\ }$. On the other side it follows easily from Proposition 1.4.7 and the continuity of inducing representations that $\ker(\mathrm{ind}_{H_\rho}^G(\rho \times U)) = \ker(\mathrm{ind}_{H_\rho}^G(\rho' \times U'))$ whenever $\rho \times U$ and $\rho' \times U'$ lie in the same quasi-orbit. □

The following theorem, which is certainly well known to the experts (see for instance [30, 28, 35]), shows that the symmetry groups play a similar role as the stabilizers in the representation theory of abelian systems. We now show that a relatively easy proof can be obtained as an application of Theorem 3.1.7. Before we state the theorem let us introduce the following notation: If (A, G, α, τ) is a separable twisted abelian system such that A is type I, then for $\rho \in \widehat{A}$, we will denote by Σ_ρ the *symmetrizer* or *symmetry group* of ρ, which is just the inverse image of the symmetry group $\tilde{\Sigma}_{\tilde\omega_\rho}$ of the Mackey obstruction $[\tilde\omega_\rho]$ of ρ under the quotient map $G \to \tilde{G}$.

THEOREM 3.1.11. *Let (A, G, α, τ) be a separable abelian twisted system such that A is type I. For ρ in $\widehat{A}$ let V be a unitary representation of Σ_ρ extending ρ to a covariant representation (ρ, V) of $(A, \Sigma_\rho, \alpha, \tau)$. Then $\ker(\mathrm{ind}_{\Sigma_\rho}^G(\rho \times V))$ is a primitive ideal of $A \rtimes_{\alpha,\tau} G$ and every primitive ideal of $A \rtimes_{\alpha,\tau} G$ can be obtained in this way.*

We need the following lemma.

LEMMA 3.1.12. *Let (A, G, α, τ) be as in Theorem 3.1.11 and for ρ in $\widehat{A}$ let H_ρ be a maximally ρ-unitary subgroup of G. Then $\ker(\mathrm{ind}_{\Sigma_\rho}^G(\rho \times W|_{\Sigma_\rho})) = \ker(\mathrm{ind}_{H_\rho}^G(\rho \times W))$ for all unitary representations W of H_ρ extending ρ to a covariant representation of $(A, H_\rho, \alpha, \tau)$.*

PROOF. We know from Proposition 1.4.9 that

$$\operatorname{ind}_{\Sigma_\rho}^{H_\rho}(\rho \times W|_{\Sigma_\rho}) = \lambda_{H_\rho/\Sigma_\rho} \otimes (\rho \times W),$$

where $\lambda_{H_\rho/\Sigma_\rho}$ denotes the left regular representation of H_ρ/Σ_ρ. Since $\lambda_{H_\rho/\Sigma_\rho}$ is weakly equivalent to $\widehat{H_\rho/\Sigma_\rho}$ we conclude that $\operatorname{ind}_{\Sigma_\rho}^{H_\rho}(\rho \times W|_{\Sigma_\rho})$ is weakly equivalent to $\widehat{H_\rho/\Sigma_\rho} \otimes (\rho \times W) = \{\rho \times \chi W; \chi \in \widehat{H_\rho/\Sigma_\rho}\}$. Since $h_{\omega_\rho}^{H_\rho}(G)$ is dense in $\widehat{H_\rho/\Sigma_\rho}$ we conclude from Lemma 3.1.10 that $\widehat{H_\rho/\Sigma_\rho} \otimes (\rho \times W)$ is contained in the quasi-orbit of $\rho \times W$ in $(A \rtimes_{\alpha,\tau} H_\rho)^\wedge$. But this implies that $\ker(\operatorname{ind}_{H_\rho}^G(\rho \times \chi W)) = \ker(\operatorname{ind}_{H_\rho}^G(\rho \times W))$ for all $\chi \in \widehat{H_\rho/\Sigma_\rho}$. Thus it follows from the continuity of induction and the theorem of induction in stages that

$$\ker(\operatorname{ind}_{\Sigma_\rho}^G(\rho \times W|_{\Sigma_\rho})) = \ker(\operatorname{ind}_{H_\rho}^G(\operatorname{ind}_{\Sigma_\rho}^{H_\rho}(\rho \times W|_{\Sigma_\rho})))$$

$$= \bigcap_{\chi \in \widehat{H_\rho/\Sigma_\rho}} \ker(\operatorname{ind}_{H_\rho}^G(\rho \times \chi W)) = \ker(\operatorname{ind}_{H_\rho}^G(\rho \times W)).$$

□

PROOF OF THEOREM 3.1.11. Let $H : \widehat{A} \to \mathfrak{K}(G)$ be a choice of maximally pointwise unitary subgroups of G. Using Lemma 3.1.12 and Theorem 3.1.7 we see that Theorem 3.1.11 follows easily from the fact that the map

$$\operatorname{res}_{\Sigma_\rho}^{H_\rho} : (A_\rho \rtimes_{\alpha,\tau} H_\rho)^\wedge \mapsto (A_\rho \rtimes_{\alpha,\tau} \Sigma_\rho)^\wedge; \rho \times W \to \rho \times W|_{\Sigma_\rho}$$

is surjective, which follows from the fact that $(A_\rho \rtimes_{\alpha,\tau} \Sigma_\rho)^\wedge$ is equal to $\{\rho \times \nu W|_{\Sigma_\rho}; \nu \in \widehat{\Sigma_\rho}\}$ if $\rho \times W$ is any element in $(A \rtimes_{\alpha,\tau} H_\rho)^\wedge$. □

In the special case where $A = \mathcal{K}$, the algebra of compact operators, we obtain the following well known but fundamental result (see [**33**, Proposition 34] and [**35**, Theorem 1.1]).

PROPOSITION 3.1.13. *Suppose that $(\mathcal{K}, G, \alpha, \tau)$ is a separable abelian twisted system, where $\mathcal{K}$ denotes the compact operators on some separable Hilbert space $\mathcal{H}$. Let Σ be the symmetry group of $(\mathcal{K}, G, \alpha, \tau)$. Then*

$$\operatorname{Ind}_\Sigma^G : (\mathcal{K} \rtimes_{\alpha,\tau} \Sigma)^\wedge \to \operatorname{Prim}(\mathcal{K} \rtimes_{\alpha,\tau} G); \rho \times V \to \ker(\operatorname{ind}_\Sigma^G(\rho \times V))$$

is a homeomorphism with inverse given by the restriction map

$$\operatorname{Res}_\Sigma^G : \operatorname{Prim}(\mathcal{K} \rtimes_{\alpha,\tau} G) \to \operatorname{Prim}(\mathcal{K} \rtimes_\alpha \Sigma) \cong (\mathcal{K} \rtimes_\alpha \Sigma)^\wedge,$$

which is defined by $\operatorname{Res}_\Sigma^G(\ker(\pi \times U)) = \ker(\pi \times U|_\Sigma)$. Moreover, for each $P \in \operatorname{Prim}(\mathcal{K} \rtimes_{\alpha,\tau} G)$, the map $\widehat{G} \to \operatorname{Prim}(\mathcal{K} \rtimes_{\alpha,\tau} G); \chi \mapsto \widehat{\alpha}_\chi(P)$ factors through a homeomorphism between $\widehat{\widetilde{\Sigma}}$ and $\operatorname{Prim}(\mathcal{K} \rtimes_{\alpha,\tau} G)$. In particular, each $P \in \operatorname{Prim}(\mathcal{K} \rtimes_{\alpha,\tau} G)$ has constant stabilizer $\Sigma^\perp$ under the action of $\widehat{G}$.

For the proof we need the following easy lemma, which follows by similar arguments as used in the proof of Lemma 3.1.10.

LEMMA 3.1.14. *Let $(\mathcal{K}, G, \alpha, \tau)$ and Σ be as in Proposition 3.1.13. Then the canonical action of G on $(\mathcal{K} \rtimes_{\alpha,\tau} \Sigma)\widehat{}$ is trivial.*

PROOF OF PROPOSITION 3.1.13. It follows from Theorem 3.1.11 that the map Ind_Σ^G is well defined and surjective. In order to see that Res_Σ^G is inverse to Ind_Σ^G let $\rho \times V \in (\mathcal{K} \rtimes_{\alpha,\tau} \Sigma)\widehat{}$. Using Proposition 1.4.8 we see that

$$\text{Res}_\Sigma^G(\text{Ind}_\Sigma^G(\rho \times V)) = \ker(\text{res}_\Sigma^G(\text{ind}_\Sigma^G(\rho \times V)) = \bigcap_{s \in G} \ker((\rho \times V) \circ \gamma_s^\Sigma) = \ker(\rho \times V)$$

since G acts trivially on $(\mathcal{K} \rtimes_{\alpha,\tau} \Sigma)\widehat{}$. The continuity of restriction and induction now implies that Ind_Σ^G is a homeomorphism.

In order to complete the proof let $P = \ker(\text{ind}_\Sigma^G(\rho \times V)) \in \text{Prim}(\mathcal{K} \rtimes_{\alpha,\tau} G)$. Then

$$\widehat{\alpha}_\chi(\ker(\text{ind}_\Sigma^G(\rho \times V))) = \ker(\chi \otimes \text{ind}_\Sigma^G(\rho \times V)) = \ker(\text{ind}_\Sigma^G(\rho \times \chi|_\Sigma V)).$$

Thus the homeomorphism Ind_Σ^G of Proposition 3.1.13 is $\widehat{\widehat{G}}$-equivariant with respect to the canonical action of $\widehat{\widehat{G}}$ on $(\mathcal{K} \rtimes_{\alpha,\tau} \Sigma)\widehat{}$, which easily completes the proof. $\square$

3.2. Abelian twisted systems with type I crossed products

Recall that a C^*-algebra B is called CCR (or liminal), if $\rho(B) = \mathcal{K}(\mathcal{H}_\rho)$ for every irreducible representation ρ of B. Note that this condition easily implies that every one-point set in $\widehat{B}$ is closed. A C^*-algebra B is called GCR (or postliminal) if $\rho(B)$ contains $\mathcal{K}(\mathcal{H}_\rho)$ for every irreducible representation ρ of B. This condition implies that all points in $\widehat{B}$ are locally closed and is equivalent to the type I condition by [**9**, 9.5.9]. Moreover, if B is separable, then B is type I (resp. CCR) if and only if all points in $\widehat{B}$ are locally closed (resp. closed) [**9**, 4.7.15 and Theorem 9.1].

It will be convenient to use the following notations: If B is a C^*-algebra and $\rho \in \widehat{B}$, then we say that ρ is a GCR *element of* $\widehat{B}$ if $\rho(B)$ contains the compact operators $\mathcal{K}(\mathcal{H}_\rho)$. Furthermore, ρ is called a CCR *element* if $\rho(B)$ is equal to $\mathcal{K}(\mathcal{H}_\rho)$. Thus, B is type I (resp. CCR) if and only if each $\rho \in \widehat{B}$ is GCR (resp. CCR). If B is separable, then it follows from [**9**, Theorem 9.1] that an element $\rho \in \widehat{B}$ is GCR (resp. CCR) if and only if $\{\rho\}$ is locally closed (resp. closed) in $\widehat{B}$. The following theorem gives some equivalent conditions for $A \rtimes_{\alpha,\tau} G$ being type I or CCR. See also [**28**] for some other results on the type I problem for crossed products by abelian groups, and [**1**, Chapter II] for related results for group C^*-algebras.

THEOREM 3.2.1. *Suppose that (A, G, α, τ) is a separable abelian twisted system such that A is type I. Let $H : \widehat{A} \to \mathfrak{K}(G); \rho \mapsto H_\rho$ be a choice of maximally pointwise unitary subgroups for (A, G, α, τ) and let $S : \widehat{A} \to \mathfrak{K}(G)$ denote the* stabilizer map. Moreover, let $\mathcal{R} = \{\rho \times U \in (A \rtimes_{\alpha,\tau} H_\rho)\widehat{\ }; \rho \in \widehat{A}\}$ be as in Theorem 3.1.7. Then the following conditions are equivalent:

(1) $A \rtimes_{\alpha,\tau} G$ is type I (resp. CCR).

(2) The G-orbit $G(\rho \times U)$ is locally closed (resp. closed) in $(A \rtimes_{\alpha,\tau} H_\rho)\widehat{\ }$ for all $\rho \times U \in \mathcal{R}$.

(3) All Mackey obstructions of (A, G, α, τ) are type I and the G-orbits $G(\sigma \times W)$ are locally closed (resp. closed) in $(A \rtimes_{\alpha,\tau} S_\rho)\widehat{\ }$ for all $\rho \in \widehat{A}$ and $\sigma \times W \in (A \rtimes_{\alpha,\tau} S_\rho)\widehat{\ }$ such that $\ker \sigma = \ker \rho$.

Moreover, if $A \rtimes_{\alpha,\tau} G$ is type I then $\mathrm{ind}_{H_\rho}^G(\rho \times U)$ is irreducible for all $\rho \times U \in \mathcal{R}$ and $\mathrm{ind}_{H_\rho}^G(\rho \times U)$ is equivalent to $\mathrm{ind}_{H_{\rho'}}^G(\rho' \times U')$ for some other $\rho' \times U' \in \mathcal{R}$, if and only if $\rho' \times U' = (\rho \times U) \circ \gamma_s^{H_\rho}$ for some $s \in G$. An analogous result is true for the representations $\sigma \times W$ of Condition (3).

The following lemma constitutes the main part of the proof of Theorem 3.2.1

LEMMA 3.2.2. Let (A, G, α, τ) be a separable abelian twisted system and let $\rho \in \widehat{A}$ such that the stabilizer of $Q = \ker \rho \in \mathrm{Prim}(A)$ is equal to N_τ. Then the following conditions are equivalent:

(1) $\ker(\mathrm{ind}_{N_\tau}^G \rho)$ is the kernel of a GCR element of $A \rtimes_{\alpha,\tau} G$.

(2) ρ is a GCR element of $\widehat{A}$ and the G-orbit $G(\rho) \subseteq \widehat{A}$ is locally closed in $\widehat{A}$.

(3) $\mathrm{ind}_{N_\tau}^G \rho$ is a GCR element of $(A \rtimes_{\alpha,\tau} G)\widehat{\ }$.

Moreover, if Conditions (1) to (3) are satisfied, then $\mathrm{ind}_{N_\tau}^G \rho$ is a CCR element of $(A \rtimes_{\alpha,\tau} G)\widehat{\ }$ if and only if $G(\rho)$ is closed in $\widehat{A}$.

PROOF. Since all conditions in the lemma are invariant under passing over to a Morita equivalent action, we may assume that τ is trivial. Assume that (1) holds and let $P = \ker(\mathrm{ind}_{\{e\}}^G \rho)$. Since for any representation $\pi \times U$ of $A \rtimes_\alpha G$ and $\chi \in \widehat{G}$ we have the relation

$$\widehat{\alpha}_\chi(\ker(\pi \times U)) = \ker(\pi \times \chi U) = \ker(\chi \otimes (\pi \times U))$$

we conclude from Proposition 1.4.9 that

$$\widehat{\alpha}_\chi(P) = \ker(\chi \otimes \mathrm{ind}_{\{e\}}^G \rho) = \ker(\mathrm{ind}_{\{e\}}^G \rho) = P$$

for all $\chi \in \widehat{G}$. Suppose now that $\pi \times U \in (A \rtimes_\alpha G)\widehat{\ }$ such that $P = \ker(\pi \times U)$. Then $\pi \times U$ is a GCR-element of $(A \rtimes_\alpha G)\widehat{\ }$ by assumption. Thus $\{\pi \times U\}$ is a $\widehat{G}$-invariant locally closed subset of $\mathrm{Prim}(A \rtimes_\alpha G)$. Let B denote the corresponding subquotient of $A \rtimes_\alpha G$. Then $B \rtimes_{\widehat{\alpha}} \widehat{G}$ is a locally closed subset of $(A \rtimes_\alpha G) \rtimes_{\widehat{\alpha}} \widehat{G}$.

We show that $B \rtimes_{\widehat{\alpha}} \widehat{G}$ corresponds to $G(\rho)$ under the canonical homeomorphism $\mathrm{IND} : \widehat{A} \to ((A \rtimes_\alpha G) \rtimes_{\widehat{\alpha}} \widehat{G})\widehat{\ }$ (see the constructions preceding Theorem 2.2.2). To this end suppose that ρ' is an element of $\widehat{A}$ such that

IND $\rho' \in (B \rtimes_{\hat{\alpha}} \widehat{G})\widehat{\;}$. Then it follows from Theorem 2.2.2 that

$$\ker(\pi \times U) = \ker(\operatorname{res}^{\widehat{G}}_{\{1_G\}}(\operatorname{IND} \rho')) = \ker(\operatorname{ind}^G_{\{e\}} \rho'),$$

from which follows that ρ' is in the quasi-orbit of ρ. But since IND is G-equivariant (with respect to the actions α and $\widehat{\widehat{\alpha}}$), and since, by Proposition 3.1.13, $\operatorname{Prim}(B \rtimes_{\hat{\alpha}} \widehat{G})$ consists of only one G-orbit with respect to the double dual action $\widehat{\widehat{\alpha}}$, we conclude that the orbit of $\ker \rho'$ in $\operatorname{Prim}(A)$ is locally closed. But this implies that $G(\ker \rho) = G(\ker \rho')$.

Now let S be a closed subgroup of G such that $S^\perp$ is the symmetry group of the Mackey obstruction of the system $(B, \widehat{G}, \widehat{\alpha})$. Then the stabilizer of $\ker(\operatorname{IND} \rho)$ is equal to S by Proposition 3.1.13. Since the stabilizer of $\ker \rho$ is trivial by assumption, it follows from the G-equivariance of IND that S is trivial, too. We conclude that the symmetry group of $(B, \widehat{G}, \ddot{\alpha})$ is all of $\widehat{G}$, thus the Mackey obstruction of the system $(B, \widehat{G}, \widehat{\alpha})$ is trivial and $(B \rtimes_{\hat{\alpha}} \widehat{G})\widehat{\;}$ is homeomorphic to G, from which now follows that $s \mapsto \rho \circ \alpha_{s^{-1}}$ is a homeomorphism of G onto $G(\rho)$. In particular, we see that ρ is a GCR element of $\widehat{A}$ and $G(\rho)$ is a locally closed subset of $\widehat{A}$.

Assume now that (2) holds. Let $A_{G(\rho)}$ denote the subquotient of A defined by the locally closed subset $G(\rho) \subseteq \widehat{A}$. Then it follows from Green's theorem (Theorem 1.4.12) that $\{\operatorname{ind}^G_{\{e\}} \rho\} = (A_{G(\rho)} \rtimes_\alpha G)\widehat{\;}$ is a locally closed subset of $(A \rtimes_\alpha G)\widehat{\;}$ from which follows that $\operatorname{ind}^G_{\{e\}} \rho$ is GCR. Thus (3) holds.

The implication (3)$\Rightarrow$(1) is trivial. The assertion about CCR elements follows from the arguments above and the fact that F is a closed G-invariant subset of $\widehat{A}$ if and only if $(A_F \rtimes_\alpha G)\widehat{\;}$ is a closed $\widehat{G}$-invariant subset of $(A \rtimes_\alpha G)\widehat{\;}$. $\square$

PROOF OF THEOREM 3.2.1. Let $\rho \times U \in \mathcal{R}$. Then, by Lemma 3.1.10, we know that the stability group of $\rho \times U$ under the action of G on $(A \rtimes_{\alpha,\tau} H_\rho)\widehat{\;}$ is equal to H_ρ. Thus, by passing over to the system $(A \rtimes_{\alpha,\tau} H_\rho, G, \gamma^{H_\rho}, \tau^{H_\rho})$, it follows from Lemma 3.2.2 and the fact that $\ker(\operatorname{ind}^G_{H_\rho}(\rho \times U))$ is a primitive ideal of $A \rtimes_{\alpha,\tau} G$ that $\ker(\operatorname{ind}^G_{H_\rho}(\rho \times U))$ is the kernel of a GCR (resp. CCR) element of $(A \rtimes_{\alpha,\tau} G)\widehat{\;}$ if and only if $G(\rho \times U)$ is locally closed (resp. closed) in $(A \rtimes_{\alpha,\tau} H_\rho)\widehat{\;}$. Using Theorem 3.1.7, this easily implies (1) $\Leftrightarrow$ (2).

In order to show (1) $\Leftrightarrow$ (3) let $\rho \in \widehat{A}$ and let $\sigma \times W \in (A \rtimes_{\alpha,\tau} S_\rho)\widehat{\;}$ such that $\ker \sigma = \ker \rho$, i.e. $\sigma \times W \in (A_\rho \rtimes_{\alpha,\tau} S_\rho)\widehat{\;}$. The stabilizer of $\ker(\sigma \times W)$ under the action of G is trivially seen to be S_ρ. Thus, by applying Lemma 3.2.2 to the system $(A \rtimes_{\alpha,\tau} S_\rho, G, \gamma^{S_\rho}, \tau^{S_\rho})$, we see that $\ker(\operatorname{ind}^G_{S_\rho}(\sigma \times W))$ is the kernel of a GCR (resp. CCR) element in $(A \rtimes_{\alpha,\tau} G)\widehat{\;}$ if and only if $\sigma \times W$ is a GCR element of $(A \rtimes_{\alpha,\tau} S_\rho)\widehat{\;}$ and $G(\sigma \times W)$ is locally closed (resp. closed) in $(A \rtimes_{\alpha,\tau} S_\rho)\widehat{\;}$. It then follows from the Gootman-Rosenberg theorem that $A \rtimes_{\alpha,\tau} G$ is type I (resp. CCR) if and only if $A_\rho \rtimes_{\alpha,\tau} S_\rho$ is type I for all ρ in $\widehat{A}$ and $G(\sigma \times W)$ is locally closed (resp. closed) for all $\sigma \times W \in (A_\rho \rtimes_{\alpha,\tau} S_\rho)\widehat{\;}$. But $A_\rho \rtimes_{\alpha,\tau} S_\rho$ is isomorphic to $C^*(\tilde{S}_\rho, \tilde{\omega}_\rho^{-1}) \otimes \mathcal{K}(\mathcal{H}_\rho)$ by Theorem 1.4.15, where $[\tilde{\omega}_\rho]$ denotes the Mackey obstruction of ρ. Furthermore, $\tilde{\omega}_\rho$ is type I if and only if $\tilde{\omega}_\rho^{-1}$ is type I,

since $h_{\tilde{\omega}_\rho}$ has image $\widehat{G/\Sigma_\rho}$ if and only if $h_{\tilde{\omega}_\rho^{-1}}$ has image $\widehat{G/\Sigma_\rho}$. It follows that
$(1) \Leftrightarrow (3)$.

Let us finally mention that, in case where $A \rtimes_{\alpha,\tau} G$ is type I, it follows also
from Lemma 3.2.2, applied to $(A \rtimes_{\alpha,\tau} H_\rho, G, \gamma^{H_\rho}, \tau^{H_\rho})$, that $\mathrm{ind}_{H_\rho}^G (\rho \times U)$ is
irreducible for all $\rho \times U \in \mathcal{R}$. The fact that $\mathrm{ind}_{H_\rho}^G (\rho \times U)$ is equivalent to
$\mathrm{ind}_{H_\rho}^G (\rho' \times U')$ (for $\rho \times U, \rho' \times U' \in \mathcal{R}$) if and only if $\rho' \times U'$ is in the G-orbit of
$\rho \times U$ follows from Theorem 3.1.7 together with the fact that $G(\rho \times U)$ is equal
to the quasi-orbit of $\rho \times U$ since $G(\rho \times U)$ is locally closed. Similar arguments
apply to the elements of $(A_\rho \rtimes_{\alpha,\tau} S_\rho)\hat{\,}$. $\quad\square$

Before we close this section, we want to give an application of the results of
the last two sections for the representation theory of locally compact groups. For
this let us assume that N is a type I (i.e. the group C^*-algebra $C^*(N)$ is type I)
closed normal subgroup of the second countable locally compact group G such
that G/N is abelian. Then $C^*(G)$ is isomorphic to the twisted crossed product
$C^*(N) \rtimes_{\gamma^N, \tau^N} G$ of the abelian twisted system $(C^*(N), G, \gamma^N, \tau^N)$. If $V \in \widehat{N}$,
then a closed subgroup H of G is maximally V-unitary with respect to the twisted
action (γ^N, τ^N) if and only if H is maximal with respect to the property that
there exists a unitary extension, say $\bar{V}$, of V to H. Thus, combining Proposition
3.1.4 with Theorem 3.1.7 and Theorem 3.2.1 we obtain the following result:

THEOREM 3.2.3. *Let G be a separable locally compact group and let N be a
closed normal type I subgroup of G such that G/N is abelian.*

(1) *There exists a map $H : \widehat{N} \to \mathfrak{K}(G); V \to H_V$ which is constant on quasi-
orbits in $\widehat{N}$ such that each H_V is a maximal closed subgroup of G with
respect to the property that there exists a unitary extension $\bar{V}$ of V to
H_V.*

(2) *Let $H : \widehat{N} \to \mathfrak{K}(G)$ be as in (1) and let $\mathcal{R} = \{W \in \widehat{H}_V; V \in \widehat{N}, W|_N = V\}$. Then $\ker(\mathrm{ind}_{H_V}^G W)$ is a primitive ideal in $C^*(G)$ for any $W \in \mathcal{R}$ and every primitive ideal can be obtained in this way. Moreover,
$\ker(\mathrm{ind}_{H_V}^G W) = \ker(\mathrm{ind}_{H_{V'}}^G W')$ if and only if $V = W|_N$ and $V' = W'|_N$
lie in the same quasi-orbit in $\widehat{N}$ and W and W' lie in the same quasi-orbit
in $\widehat{H}_V$.*

(3) *G is type I (resp. CCR) if and only if the G-orbit $G(W)$ is locally closed
(resp. closed) in $\widehat{H}_V$ for all $W \in \mathcal{R}$. If this is true, then $\mathrm{ind}_{H_V}^G W$ is
irreducible for all $W \in \mathcal{R}$, and induction defines a bijection between the
G-orbits in $\mathcal{R}$ and $\widehat{G}$.*

A locally compact group G is called *monomial* if every irreducible representation of G is induced from a one-dimensional representation of some subgroup.
Recall also that a group G is called *metabelian* (or two-step solvable) if the
commutator subgroup $[G, G]$ of G is abelian.

COROLLARY 3.2.4. *Every separable type I metabelian group G is monomial.*

PROOF. Apply Theorem 3.2.3 to the closure N of $[G, G]$. $\square$

3.3. Maximally pointwise unitary subgroups for dual systems

If (A, G, α, τ) is a separable twisted abelian system such that A is type I, then we saw that there are three different maps from $\widehat{A} \to \mathfrak{K}(G)$, all constant on quasi-orbits, which play an important role in the representation theory of the system. These maps are the stabilizer map

$$S : \widehat{A} \to \mathfrak{K}(G); \rho \mapsto S_\rho,$$

the symmetrizer map

$$\Sigma : \widehat{A} \to \mathfrak{K}(G); \rho \mapsto \Sigma_\rho,$$

where Σ_ρ denotes the pull-back of the symmetry group $\tilde{\Sigma}_\rho \subseteq \tilde{G} = G/N_\tau$ of the Mackey obstruction $[\tilde{\omega}_\rho] \in H^2(\tilde{S}_\rho, \mathbb{T})$ of ρ, and finally a choice

$$H : \widehat{A} \to \mathfrak{K}(G); \rho \mapsto H_\rho$$

of maximally pointwise unitary subgroups for (A, G, α, τ). Recall also that the maps S and Σ are uniquely determined by the system (A, G, α, τ), while H is in general really a choice, i.e., there may be different possibilities for defining H_ρ. If (α, τ) is pointwise unitary on the stabilizers, i.e., if all Mackey obstructions of (A, G, α, τ) vanish, then the maps S, Σ and H automatically coincide.

Suppose now that (A, G, α, τ) is as above, and that in addition $A \rtimes_{\alpha, \tau} G$ is type I. Then we also have a stabilizer map, a symmetrizer map and choices of maximally pointwise unitary subgroups for the dual system $(A \rtimes_{\alpha, \tau} G, \widehat{\widehat{G}}, \widehat{\alpha})$. We are now going to relate these maps with the original maps S, Σ and H for (A, G, α, τ). Having this done, we will give a relation between the Mackey obstructions of the dual system $(A \rtimes_{\alpha, \tau} G, \widehat{G}, \widehat{\alpha})$ and the Mackey obstructions of (A, G, α, τ).

Since all maps given above are constant on quasi-orbits we start with relating the quasi-orbit space $\mathcal{Q}_G(\widehat{A})$ of $\widehat{A}$ with the quasi-orbit space $\mathcal{Q}_{\widehat{G}}((A \rtimes_{\alpha, \tau} G)^\smallfrown)$. For this recall that $\mathrm{Res}^G : (A \rtimes_{\alpha, \tau} G)^\smallfrown \to \mathcal{Q}_G(\widehat{A})$ denotes the quasi-orbit map which maps each $\pi \times U \in (A \rtimes_{\alpha, \tau} G)^\smallfrown$ onto the unique quasi-orbit $\mathcal{Q}(\rho)$ such that $\ker \pi = \ker \mathcal{Q}(\rho)$. The following proposition was first proved by Gootman and Lazar [**29**, Corollary 2.5]. For the readers convenience, and since we need some of the constructions later in this work, we include a slightly different proof here.

PROPOSITION 3.3.1. *Suppose that (A, G, α, τ) is a separable twisted abelian system. Then the quasi-orbit map $\mathrm{Res}^G : (A \rtimes_{\alpha, \tau} G)^\smallfrown \to \mathcal{Q}_G(\widehat{A})$ is constant on $\widehat{G}$-quasi-orbits in $(A \rtimes_{\alpha, \tau} G)^\smallfrown$ and factors through a homeomorphism between $\mathcal{Q}_G(\widehat{A})$ and $\mathcal{Q}_{\widehat{G}}((A \rtimes_{\alpha, \tau} G)^\smallfrown)$.*

PROOF. Since $\pi \times U \circ \widehat{\alpha}_\chi = \pi \times \overline{\chi} U$, it follows directly from the definition and the continuity of Res^G, that Res^G is constant on $\widehat{\widetilde{G}}$-quasi-orbits in $(A \rtimes_{\alpha,\tau} G)\widehat{}$. Let $\mathrm{IND} : \widehat{A} \to ((A \rtimes_{\alpha,\tau} G) \rtimes_{\widehat{\alpha}} \widetilde{G})\widehat{}$ be the canonical homeomorphism as defined in Section 2.2. Then $\ker(\mathrm{ind}_{N_\tau}^G \rho) = \ker(\mathrm{res}_{\{1_G\}}^{\widehat{\widetilde{G}}}(\mathrm{IND}\,\rho))$ for any $\rho \in \widehat{A}$ by Theorem 2.2.2. Thus, by Proposition 1.4.13, there exists a unique quasi-orbit $\mathcal{Q}(\pi \times U)$ in $\mathcal{Q}_{\widehat{\widetilde{G}}}((A \rtimes_{\alpha,\tau} G)\widehat{})$ such that $\ker \mathcal{Q}(\pi \times U) = \ker(\mathrm{ind}_{N_\tau}^G \rho)$. If we denote this quasi-orbit by $\mathrm{Ind}^G(\rho)$, we obtain therefore a continuous map $\mathrm{Ind}^G : \widehat{A} \to \mathcal{Q}_{\widehat{\widetilde{G}}}((A \rtimes_{\alpha,\tau} G)\widehat{})$ which is constant on G-quasi-orbits by the continuity of induction. Hence, in order to complete the proof, it remains to show that Ind^G is inverse to Res^G. To this end observe that $\ker(\mathrm{ind}_{N_\tau}^G \rho) = \bigcap\{\ker(\pi \times U); \pi \times U \in \mathrm{Ind}^G(\rho)\}$. Since Res^G is constant on quasi-orbits we obtain from the continuity of restriction and from Proposition 1.4.8 that

$$\ker \pi = \bigcap\{\ker \pi'; \pi' \times U' \in \mathrm{Ind}^G(\rho)\} = \ker(\mathrm{res}_{N_\tau}^G(\mathrm{ind}_{N_\tau}^G \rho))$$

$$= \bigcap_{s \in G} \ker(\rho \circ \alpha_s) = \ker \mathcal{Q}(\rho).$$

Thus we see that $\mathrm{Res}^G \circ \mathrm{Ind}^G$ is the identity on $\mathcal{Q}_G(\widehat{A})$. Using duality, the same arguments imply that $\mathrm{Ind}^G \circ \mathrm{Res}^G$ is the identity on $\mathcal{Q}_{\widehat{\widetilde{G}}}((A \rtimes_{\alpha,\tau} G)\widehat{})$. $\square$

We proceed with a characterization of maximally pointwise unitary subgroups of G in case where $A \rtimes_{\alpha,\tau} G$ is type I.

LEMMA 3.3.2. *Suppose that (A, G, α, τ) is a separable twisted abelian system such that A and $A \rtimes_{\alpha,\tau} G$ are type I. Let $\rho \in \widehat{A}$. Then a subgroup H of G is maximally ρ-unitary if and only if there exists a unitary representation U of H extending ρ to a covariant representation (ρ, U) of (A, H, α, τ) such that $\mathrm{ind}_H^G(\rho \times U)$ is irreducible.*

PROOF. It follows from Theorem 3.2.1 and the definition of maximally ρ-unitary subgroups of G that the conditions in the lemma are necessary for H being maximally ρ-unitary. In order to see that these conditions are also sufficient, assume that L is a maximally ρ-unitary subgroup of G containing H. Then there exists a unitary representation V of L extending ρ to a covariant representation of (A, L, α, τ). By multiplying V with a suitable character of L we may assume that $V|_H = U$. Thus, using Proposition 1.4.9, we see that

$$\mathrm{ind}_H^G(\rho \times U) = \mathrm{ind}_L^G(\mathrm{ind}_H^L(\rho \times V|_H)) = \mathrm{ind}_L^G(\lambda_{L/H} \otimes (\rho \times V)),$$

where $\lambda_{L/H}$ denotes the left regular representation of L/H. If $L \neq H$, then $\lambda_{L/H} \otimes (\rho \times V)$ is not irreducible, which implies that $\mathrm{ind}_H^G(\rho \times U) = \mathrm{ind}_L^G(\lambda_{L/H} \otimes (\rho \times V))$ is also not irreducible. But this contradicts the assumptions. $\square$

THEOREM 3.3.3. *Suppose that (A, G, α, τ) is a separable abelian twisted system such that A and $A \rtimes_{\alpha,\tau} G$ are type I. Let*

$$S, \Sigma, H : \mathcal{Q}_G(\widehat{A}) \to \mathfrak{K}(G)$$

denote the stabilizer map, the symmetrizer map and a choice of maximally pointwise unitary subgroups of G, respectively, viewed as maps on the quasi-orbit space $\mathcal{Q}_G(\widehat{A})$. Define

$$\Sigma^{\perp}, S^{\perp}, H^{\perp} : (A \rtimes_{\alpha,\tau} G)\widehat{} \to \mathfrak{K}(\widehat{\widetilde{G}})$$

by $\Sigma^{\perp}_{\pi \times U} = (\Sigma_{\mathrm{Res}^G(\pi \times U)})^{\perp}$, $S^{\perp}_{\pi \times U} = (S_{\mathrm{Res}^G(\pi \times U)})^{\perp}$ and $H^{\perp}_{\pi \times U} = (H_{\mathrm{Res}^G(\pi \times U)})^{\perp}$. Then $\Sigma^{\perp}$ is the stabilizer map, $S^{\perp}$ is the symmetrizer map and $H^{\perp}$ is a choice of maximally pointwise unitary subgroups for $(A \rtimes_{\alpha,\tau} G, \widehat{\widetilde{G}}, \widehat{\alpha})$.

PROOF. Since the dual actions of Morita equivalent actions are also Morita equivalent by [**15**, Theorem 2], we may pass over to a Morita equivalent action in order to assume that τ is trivial. Let $\pi \times U$ in $\widehat{A}$, and let $\rho \in \widehat{A}$ with $\mathrm{Res}^G(\pi \times U) = \mathcal{Q}(\rho)$. By Theorem 3.1.11 we know that there exists a representation V of Σ_ρ such that $\ker(\pi \times U) = \ker(\mathrm{ind}^G_{\Sigma_\rho}(\rho \times V))$. It follows then from Proposition 1.4.9 that

$$\widehat{\alpha}_\chi(\ker(\pi \times U)) = \ker(\mathrm{ind}^G_{\Sigma_\rho}(\rho \times \chi|_{\Sigma_\rho} V))$$

for all $\chi \in \widehat{G}$, from which in particular follows that $(\Sigma_\rho)^{\perp}$ is contained in the stabilizer of $\ker(\pi \times U)$, and hence also in the stabilizer of $\pi \times U$ since $A \rtimes_\alpha G$ is type I. Thus, in order to see that $(\Sigma_\rho)^{\perp}$ is equal to the stabilizer of $\pi \times U$, we have to show that $\ker(\mathrm{ind}^G_{\Sigma_\rho}(\rho \times \nu V)) \neq \ker(\mathrm{ind}^G_{\Sigma_\rho}(\rho \times V))$ if $\nu \in \widehat{\Sigma}_\rho$ is not trivial.

For this let W be a unitary representation of H_ρ which extends ρ to a covariant representation of (A, H_ρ, α) such that $W|_{\Sigma_\rho} = V$ (the last equation can be obtained by multiplying W with a suitable character of H_ρ). Choose $\chi \in \widehat{H}_\rho$ such that $\chi|_{\Sigma_\rho} = \nu$. Then it follows from Lemma 3.1.12 that $\ker(\mathrm{ind}^G_{\Sigma_\rho}(\rho \times V)) = \ker(\mathrm{ind}^G_{H_\rho}(\rho \times W))$ and $\ker(\mathrm{ind}^G_{\Sigma_\rho}(\rho \times \nu V)) = \ker(\mathrm{ind}^G_{H_\rho}(\rho \times \chi W))$. Thus, combining Theorem 3.1.7 with Theorem 3.2.1, it is enough to show that $\rho \times \chi W$ is not contained in the G-orbit of $\rho \times W$ in $(A \rtimes_\alpha H_\rho)\widehat{}$. If $s \notin S_\rho$, then $(\rho \times W) \circ \gamma_s^{H_\rho} = \rho \circ \alpha_s \times W$, which cannot be equivalent to $\rho \times \chi W$ since $\rho \circ \alpha_s$ is not equivalent to ρ. If $s \in S_\rho$, then $(\rho \times W) \circ \gamma_s^{H_\rho}$ is equivalent to $\rho \times h^{H_\rho}_{\omega_\rho}(s)W$ by Lemma 3.1.10. Since $h^{H_\rho}_{\omega_\rho}(S_\rho) \subseteq \widehat{H_\rho/\Sigma_\rho}$, it follows that $h^{H_\rho}_{\omega_\rho}(s)|_{\Sigma_\rho}$ is trivial for all $s \in S_\rho$. But this implies that $h^{H_\rho}_{\omega_\rho}(s) \neq \chi$ for all $s \in S_\rho$, since $\nu = \chi|_{\Sigma_\rho}$ is not trivial by assumption. Thus we conclude that $(\Sigma_\rho)^{\perp}$ is the stabilizer of $\pi \times U$.

Assume now that L is a subgroup of G such that $L^{\perp}$ is the symmetrizer of $\pi \times U$. Applying the same arguments as above to the dual system $(A \rtimes_\alpha G, \widehat{G}, \widehat{\alpha})$, we see that the stabilizer of $\mathrm{IND}\,\rho \in (A \rtimes_\alpha G) \rtimes_{\widehat{\alpha}} \widehat{G}$ is equal to L, since by the arguments used in the proof of Proposition 3.3.1 $\mathrm{Res}^{\widehat{G}}(\mathrm{IND}\,\rho) = \mathrm{Ind}^G \rho$ is the quasi-orbit of $\pi \times U$. Since $\mathrm{IND} : \widehat{A} \to (A \rtimes_\alpha G) \rtimes_{\widehat{\alpha}} \widehat{G}$ is G-equivariant, it follows that $L = S_\rho$. Thus, $(S_\rho)^\perp$ is the symmetrizer of $\pi \times U$.

Finally, let W be a unitary representation of H_ρ such that $\pi \times U = \mathrm{ind}_{H_\rho}^G (\rho \times W)$ (see Theorem 3.2.1). Let $\mathrm{IND}^{H_\rho} : (A \rtimes_\alpha H_\rho)\widehat{\ } \to ((A \rtimes_\alpha G) \rtimes_{\widehat{\alpha}} H_\rho^\perp)\widehat{\ }$ denote the canonical homeomorphism as described in Section 2.2. By Theorem 2.2.2 we have

$$\mathrm{res}_{\{1_G\}}^{H_\rho^\perp}(\mathrm{IND}^{H_\rho}(\rho \times W)) = \mathrm{ind}_{H_\rho}^G(\rho \times W) = \pi \times U$$

and

$$\mathrm{ind}_{H_\rho^\perp}^{\widehat{G}}(\mathrm{IND}^{H_\rho}(\rho \times W)) = \mathrm{IND}(\mathrm{res}_{\{e\}}^{H_\rho}(\rho \times W)) = \mathrm{IND}\,\rho.$$

Since $\mathrm{IND}\,\rho$ is irreducible, it follows now from Lemma 3.3.2 that $(H_\rho)^\perp$ is a maximally $\pi \times U$-unitary subgroup of $\widehat{G}$. This completes the proof. $\square$

We are now going to describe the Mackey obstructions of the dual system $(A \rtimes_{\alpha,\tau} G, \widehat{G}, \widehat{\alpha})$ in terms of the Mackey obstructions of (A, G, α, τ). Recall again that if ω is a totally skew multiplier on the abelian group G, i.e., if the symmetry group Σ_ω is trivial, then ω is a type I multiplier if and only if $h_\omega : G \to \widehat{G}$ is an isomorphism of locally compact groups [3, Theorem 3.2].

DEFINITION 3.3.4. Let ω be a type I totally skew multiplier on the separable abelian group G. Then we define the *dual multiplier* $\widehat{\omega}$ of ω on $\widehat{G}$ by

$$\widehat{\omega}(\chi, \mu) = \omega(h_\omega^{-1}(\chi), h_\omega^{-1}(\mu))^{-1}$$

for all $\chi, \mu \in \widehat{G}$.

LEMMA 3.3.5. *Let $(\mathcal{K}, G, \alpha)$ be a separable abelian system, with $\mathcal{K}$ being the compact operators on some Hilbert space $\mathcal{H}$. Assume that the Mackey obstruction $[\omega]$ of $(\mathcal{K}, G, \alpha)$ is type I and totally skew. Then $(\mathcal{K} \rtimes_\alpha G)\widehat{\ }$ consists of a single point and the Mackey obstruction of $(\mathcal{K} \rtimes_\alpha G, \widehat{G}, \widehat{\alpha})$ is equal to $[\widehat{\omega}] \in H^2(\widehat{G}, \mathbb{T})$.*

PROOF. The fact that $(\mathcal{K} \rtimes_\alpha G)\widehat{\ }$ consists of a single point is a direct consequence of Proposition 3.1.13 together with the type I assumption on ω. Let L be an ω-representation of G extending the identity on $\mathcal{K}$ to an ω-covariant representation (id, L) of $(\mathcal{K}, G, \alpha)$ and let V be an irreducible ω^{-1}-representation of G. Then (the equivalence class of) $(1 \otimes \mathrm{id}) \times (V \otimes L)$ is the unique element of $(\mathcal{K} \rtimes_\alpha G)\widehat{\ }$. We define $\widehat{V}_\chi = V_{h_\omega^{-1}(\chi)}$ for all $\chi \in \widehat{G}$.

It is straightforward to see that $\widehat{V}$ is a $\widehat{\omega}$-representation of $\widehat{G}$. Now let $f \in$

$C_c(G, \mathcal{K})$, and let $\chi = h_\omega(s) \in \widehat{G}$. Then, using Lemma 3.1.9, we obtain

$$(\widehat{V}_\chi \otimes 1)((1 \otimes \mathrm{id}) \times (V \otimes L)(f))(\widehat{V}_\chi^* \otimes 1)$$

$$= (V_s \otimes 1) \left(\int_G V_t \otimes f(t) L_t \, dt \right) (V_s^* \otimes 1)$$

$$= \int_G V_s V_t V_s^* \otimes f(t) L_t \, dt = \int_G V_t \otimes \overline{h_\omega(s)(t)} f(t) L_t \, dt$$

$$= \int_G V_t \otimes \overline{\chi(t)} f(t) L_t \, dt = (1 \otimes \mathrm{id}) \times (V \otimes L)(\widehat{\alpha}_\chi(f)).$$

Thus we see that $((1 \otimes \mathrm{id}) \times (V \otimes L), \widehat{V} \otimes 1)$ is a $\widehat{\omega}$-representation of $(\mathcal{K} \rtimes_\alpha G, \widehat{G}, \widehat{\alpha})$, which implies that $[\widehat{\omega}]$ is the Mackey obstruction of this system. $\square$

In order to describe the Mackey obstructions for the dual systems of more general abelian twisted systems, recall that any multiplier ω on the abelian group G is similar to a multiplier which is lifted from a totally skew multiplier on G/Σ_ω (see for instance [3, Theorem 3.1]). Thus we may assume that $\omega(sn, tm) = \omega(s, t)$ for all $s, t \in G$ and $n, m \in \Sigma_\omega$, and we may view $[\omega]$ as an element of the subgroup $H^2(G/\Sigma_\omega, \mathbb{T})$ of $H^2(G, \mathbb{T})$.

Let (A, G, α, τ) be a separable abelian twisted system such that A and $A \rtimes_{\alpha, \tau} G$ are type I. Let $\pi \times U$ in $(A \rtimes_{\alpha, \tau} G)\widehat{\ }$, and let $\rho \in \widehat{A}$ such that the quasi-orbit $\mathcal{Q}(\rho)$ of ρ is equal to $\mathrm{Res}^G(\pi \times U)$. Let $[\omega] \in H^2(S_\rho/\Sigma_\rho, \mathbb{T})$ be the Mackey obstruction of ρ. Since ω must be type I by Theorem 3.2.1, we may define the multiplier $\widehat{\omega}$ of $\widehat{S_\rho/\Sigma_\rho} = \Sigma_{\pi \times U}^\perp / S_{\pi \times U}^\perp$ as in Definition 3.3.4 (see Theorem 3.3.3 for the definitions of $\Sigma_{\pi \times U}^\perp$ and $S_{\pi \times U}^\perp$).

THEOREM 3.3.6. *Let* (A, G, α, τ), $\pi \times U$, *and* ρ *be as above. Let* $[\omega]$ *be the Mackey obstruction at* ρ, *viewed as an element of* $H^2(S_\rho/\Sigma_\rho, \mathbb{T})$, *and let* $[\widehat{\omega}] \in H^2(\Sigma_{\pi \times U}^\perp / S_{\pi \times U}^\perp, \mathbb{T})$ *be as above. Then* $[\widehat{\omega}]$ *is the Mackey obstruction of* $\pi \times U$ *with respect to the dual action* $\widehat{\alpha}$ *of* $\widehat{G}$ *on* $A \rtimes_{\alpha, \tau} G$.

We start the proof with the following lemma.

LEMMA 3.3.7. *Let* (A, G, α, τ) *be a separable abelian twisted system such that* A *is type I. For* ρ *in* $\widehat{A}$ *let* $[\omega_\rho] \in H^2(S_\rho/\Sigma_\rho, \mathbb{T})$ *be the Mackey obstruction of* ρ *and let* $\rho \times V \in (A \rtimes_{\alpha, \tau} \Sigma_\rho)\widehat{\ }$. *Then* $[\omega_\rho]$ *is equal to the Mackey obstruction of* $\rho \times V$ *with respect to the decomposition system* $(A \rtimes_{\alpha, \tau} S_\rho, G, \gamma^{\Sigma_\rho}, \tau^{\Sigma_\rho})$.

PROOF. As usual we pass over to a Morita equivalent action in order to assume that τ is trivial. It follows easily from Lemma 3.1.14 that the stabilizer of $\rho \times V$ is equal to S_ρ. Thus we may restrict our attention to the action of S_ρ on $A_\rho \rtimes_\alpha \Sigma_\rho$. Let R be an ω_ρ-representation of S_ρ such that (ρ, R) becomes an ω_ρ-covariant representation of (A, S_ρ, α). By multiplying R with a suitable character we may assume that $R|_{\Sigma_\rho} = V$. Note that R_s commutes with V_t for all $s \in S_\rho$ and $t \in \Sigma_\rho$ since

$$V_t R_s = \omega_\rho(t, s) R_{ts} = \omega_\rho(t, s) \omega_\rho(s, t)^{-1} R_s V_t = R_s V_t.$$

We show that $(\rho \times V, R)$ is a ω_ρ-covariant representation of $(A \rtimes_\alpha \Sigma_\rho, S_\rho, \gamma^{\Sigma_\rho}, \tau^{\Sigma_\rho})$. To this end let $f \in C_c(\Sigma_\rho, A)$ and $s \in S_\rho$. Then we compute

$$R_s(\rho \times V)(f))R_s^* = R_s \left(\int_{\Sigma_\rho} \rho(f(t)) V_t \, d_{\Sigma_\rho} t \right) R_s^*$$

$$= \int_{\Sigma_\rho} R_s \rho(f(t)) R_s^* V_t \, d_{\Sigma_\rho} t = \rho \circ \alpha_s \times V(f) = (\rho \times V) \circ \gamma_s^\Sigma(f).$$

Since $V = R|_{\Sigma_\rho}$, it follows also that $R_s = \rho \times V(\tau_s^{\Sigma_\rho})$ for all $s \in \Sigma_\rho$. This completes the proof. $\square$

PROOF OF THEOREM 3.3.6. Let $\pi \times U$ and ρ be as in the theorem. Let H_ρ be a maximal ρ-unitary subgroup of G, and let W be a unitary representation of H_ρ on $\mathcal{H}_\rho$ such that $\pi \times U = \text{ind}_{H_\rho}^G(\rho \times W)$, which exists by Theorem 3.2.1. Let $\sigma \times \tilde{U} = \text{ind}_{H_\rho}^{S_\rho}(\rho \times W)$. Then $\sigma \times \tilde{U} \in (A_\rho \rtimes_{\alpha,\tau} S_\rho)\hat{}$ and $\text{ind}_{S_\rho}^G(\sigma \times \tilde{U}) = \pi \times U$.

Now let $V = W|_{\Sigma_\rho}$ and let $(\gamma^{\Sigma_\rho}, \tau^{\Sigma_\rho})$ be the decomposition twisted action of S_ρ on $A_\rho \rtimes_{\alpha,\tau} \Sigma_\rho$. We know from the previous lemma that the Mackey obstruction of $\rho \times V$ is equal to $[\omega]$. Moreover, it follows from Proposition 1.4.8 that

$$\ker(\sigma \times \tilde{U}|_{\Sigma_\rho}) = \bigcap_{s \in S_\rho} \ker((\rho \times W|_{\Sigma_\rho}) \circ \gamma_s^{\Sigma_\rho}) = \ker(\rho \times V),$$

since S_ρ acts trivially on $\rho \times V$. Thus we conclude that $(\sigma \times \tilde{U}|_{\Sigma_\rho}) \times \tilde{U}$ is in the equivalence class of the single element of $((A_\rho \rtimes_\alpha \Sigma_\rho)_{\{\rho \times V\}} \rtimes_{\gamma^{\Sigma_\rho}, \tau^{\Sigma_\rho}} S_\rho)\hat{}$. Thus, by passing over to a Morita equivalent action, we conclude from Lemma 3.3.5 that $[\hat{\omega}]$ is the class of the Mackey obstruction of $\sigma \times \tilde{U}$ with respect to the dual action $\widehat{\alpha|_{S_\rho}}$ of $\widehat{S_\rho}$ on $A \rtimes_\alpha S_\rho$ (note that the canonical isomorphism between $(A \rtimes_{\alpha,\tau} \Sigma_\rho) \rtimes_{\gamma^{\Sigma_\rho}, \tau^{\Sigma_\rho}} S_\rho$ and $A \rtimes_{\alpha,\tau} S_\rho$ is $\widehat{S_\rho/\Sigma_\rho}$- equivariant). By Proposition 2.2.1 we know that $\widehat{\alpha|_{S_\rho}}$ is Morita equivalent to the decomposition twisted action $(\gamma^{S_\rho^\perp}, \tau^{S_\rho^\perp})$ of $\widehat{G}$ on $(A \rtimes_\alpha G) \rtimes_{\widehat{\alpha}} S_\rho^\perp$. Moreover, since $\text{ind}_{S_\rho}^G(\sigma \times \tilde{U}) = \pi \times U$, it follows from Theorem 2.2.2 that the restriction of $\text{IND}^{S_\rho}(\sigma \times \tilde{U})$ to $A \rtimes_\alpha G$ also coincides with $\pi \times U$. Thus it follows from Lemma 3.3.7 that the Mackey obstruction of $\pi \times U$ is equal to the Mackey obstruction of $\text{IND}^{S_\rho}(\sigma \times \tilde{U})$ with respect to $(\gamma^{S_\rho^\perp}, \tau^{S_\rho^\perp})$, which in turn is equal to $[\hat{\omega}]$ since Morita equivalent actions have the same Mackey obstructions by Proposition 2.1.5. $\square$

CHAPTER 4

Subgroup crossed products

In this chapter we are going to recall the basic definitions of subgroup algebras for twisted covariant systems as introduced in [**12, 13, 56**] and prove a couple of new results for subgroup algebras which will be needed in the investigation of crossed products with continuous trace.

4.1. Subgroup algebras and regularizations

If G is a locally compact group, then, as in the previous chapter, we denote by $\mathfrak{K}(G)$ the set of all closed subgroups of G. In [**21**] Fell defined a topology on $\mathfrak{K}(G)$, which makes $\mathfrak{K}(G)$ into a compact Hausdorff space. A base for this topology is given by the sets

$$U(\mathcal{F}, C) = \{L \in \mathfrak{K}(G); L \cap V \neq \emptyset \text{ for all } V \in \mathcal{F} \text{ and } L \cap C = \emptyset\},$$

where $\mathcal{F}$ runs through all finite families of open subsets of G and C runs through the compact subsets of G. If G is second countable, then $\mathfrak{K}(G)$ is second countable, too. The following useful description of convergence of nets in $\mathfrak{K}(G)$ follows easily from the definition of the topology.

PROPOSITION 4.1.1. *Let $(H_i)_{i \in I}$ be a net in $\mathfrak{K}(G)$ which converges to $H \in \mathfrak{K}(G)$ and let $s \in G$. Then $s \in H$ if and only if there exists a subnet $(H_{i_j})_{j \in J}$ of $(H_i)_{i \in I}$ and elements $s_j \in H_{i_j}$ such that $s_j \to s$ in G.*

A *smooth choice of Haar measures* on $\mathfrak{K}(G)$ is a choice of Haar measures $(d_L)_{L \in \mathfrak{K}(G)}$ such that the maps $\mathfrak{K}(G) \to \mathbb{C}; L \mapsto \int_L f(s) \, d_L s$ are continuous for all $f \in C_c(G)$. Smooth choices of Haar measures always exist [**27**, Appendix]. If N is a closed normal subgroup of G and $L : \Omega \to \mathfrak{K}(G); x \mapsto L_x$ is any map such that $L_x \supseteq N$ for all $x \in L$, then it is easily seen that L is continuous if and only if the map $\tilde{L} : \Omega \to \mathfrak{K}(\tilde{G}); \tilde{L}_x = L_x/N$ is continuous, where $\tilde{G} = G/N$. Moreover, if $(d_L)_{L \in \mathfrak{K}(G)}$ is a smooth choice of Haar measures on $\mathfrak{K}(G)$, then a smooth choice of Haar measures on $\mathfrak{K}(\tilde{G})$ is given by choosing Haar measures on $\tilde{L} = L/N$ such that

$$\int_{\tilde{L}} \int_N g(sn) d_N n \, d_{\tilde{L}} \tilde{s} = \int_L g(s) \, d_L s$$

for all $g \in C_c(G)$.

Assume now that (A, G, α, τ) is a twisted covariant system and let $L : \Omega \to \mathfrak{K}(G); x \mapsto L_x$ be a continuous map such that $L_x \supseteq N_\tau$ for all $x \in \Omega$. Define

$$\Omega^L := \{(x, s) \in \Omega \times G; s \in L_x\}.$$

Then it follows trivially from Proposition 4.1.1 that Ω^L is a closed subset of $\Omega \times G$. Denote by $C_c(\Omega^L, A, \tau)$ the set of all continuous A-valued functions f on Ω^L such that

$$f(x, ns) = f(x, s)\tau_{n^{-1}} \quad \text{for all } (x, s) \in \Omega^L \text{ and } n \in N_\tau,$$

and such that $(x, \tilde{s}) \mapsto \|f(x, s)\|$ has compact support in $\Omega^{\tilde{L}}$. We define convolution, involution and norm on $C_c(\Omega^L, A, \tau)$ by

$$f * g(x, s) = \int_{\tilde{L}_x} f(x, r)\alpha_r(g(x, r^{-1}s)) \, d_{\tilde{L}_x} r$$

$$f^*(x, s) = \Delta_{\tilde{L}_x}(\tilde{s}^{-1})\alpha_s(f(x, s^{-1})^*)$$

and

$$\|f\|_1 = \sup_{x \in \Omega} \int_{\tilde{L}_x} \|f(x, s)\| \, d_{\tilde{L}_x} \tilde{s}$$

with respect to a smooth choice of Haar measures on $\mathfrak{K}(\tilde{G})$ as given above. The completion $L^1(\Omega^L, A, \tau)$ of $C_c(\Omega^L, A, \tau)$ becomes a Banach $*$-algebra and we define the Ω^L-*subgroup algebra* $C^*(\Omega^L, A, \alpha, \tau)$ of (A, G, α, τ) to be the enveloping C^*-algebra of $L^1(\Omega^L, A, \tau)$ (see [12] for more details of the construction).

If $\rho \times V$ is a representation of $A \rtimes_{\alpha,\tau} L_x$, for some $x \in \Omega$, then we may lift $\rho \times V$ to a representation, denoted $(x, \rho \times V)$, of $C^*(\Omega^L, A, \alpha, \tau)$ by

$$(x, \rho \times V)(f) = \rho \times V(f(x, \cdot)), \quad f \in C_c(\Omega^L, A, \tau).$$

Such representations are called *subgroup representations* of $C^*(\Omega^L, A, \alpha, \tau)$. The collection of all subgroup representations (with dimension bounded by some cardinal $\aleph$), equipped with the relative topology from $\mathrm{Rep}(C^*(\Omega^L, A, \alpha, \tau))$, is denoted by $\mathcal{S}(\Omega^L, A, \alpha, \tau)$. All irreducible representations of $C^*(\Omega^L, A, \alpha, \tau)$ are subgroup representations, i.e. we may write

$$C^*(\Omega^L, A, \alpha, \tau)\widehat{} = \{(x, \rho \times V); x \in \Omega \text{ and } \rho \times V \in (A \rtimes_{\alpha,\tau} L_x)\widehat{}\}.$$

In the special case where $\Omega = \mathfrak{K}(\tilde{G})$ and $\mathrm{Id} : \mathfrak{K}(\tilde{G}) \to \mathfrak{K}(G)$ denotes the natural inclusion, we see that $\mathcal{S}(\mathfrak{K}(\tilde{G})^{\mathrm{Id}}, A, \alpha, \tau)$ consists of the collection of all pairs $(L, \rho \times V)$, where $L \supseteq N_\tau$ is a closed subgroup of G and $\rho \times V$ is a representation of $A \rtimes_{\alpha,\tau} L$. If $A = \mathbb{C}$ and τ is trivial, then we denote $C^*(\Omega^L, \mathbb{C}, \mathrm{id}, 1)$ simply by $C^*(\Omega^L)$. In case $\Omega = \mathfrak{K}(G)$ and Id being the identity we obtain the subgroup algebra $C^*(\mathfrak{K}(G)^{\mathrm{Id}})$ which is identical to the subgroup algebra as defined by Fell in [24]. The following proposition will be used frequently (for the proof see [12, Proposition 3]).

PROPOSITION 4.1.2. *Let $C^*(\Omega^L, A, \alpha, \tau)$ be a subgroup algebra of the twisted covariant system (A, G, α, τ) and let Λ be a closed subset of Ω. Then the natural inclusion of $C_c((\Omega \setminus \Lambda)^L, A, \tau)$ in $C_c(\Omega^L, A, \tau)$, and the restriction map from $C_c(\Omega^L, A, \tau)$ to $C_c(\Lambda^L, A, \tau)$ extend to the full subgroup algebras and define a short exact sequence*

$$0 \to C^*((\Omega \setminus \Lambda)^L, A, \alpha, \tau) \to C^*(\Omega^L, A, \alpha, \tau) \to C^*(\Lambda^L, A, \alpha, \tau) \to 0.$$

In particular, the natural projection from $\mathcal{S}(\Omega^L, A, \alpha, \tau)$ onto Ω is continuous.

Let us mention that in case where $\tilde{L}_x$ is amenable for all $x \in \Omega$, the projection from $C^*(\Omega^L, A, \alpha, \tau)\hat{\ }$ onto Ω is also open (this will be a special case of Proposition 4.2.5 below). The following proposition is [**12**, Corollary 2].

PROPOSITION 4.1.3. *Let (A, G, α, τ) be a twisted covariant system and suppose that $L^i : \Omega_i \to \mathfrak{K}(G)$ are continuous maps, $i = 1, 2$, such that $N_\tau \subseteq L_x^i$ for all $x \in \Omega_i$. Then the maps*

$$\mathrm{ind}_{L_{x_1}^1}^{L_{x_2}^2} : \{(x_2, (x_1, \rho \times V)) \in \Omega_2 \times \mathcal{S}(\Omega_1^{L^1}, A, \alpha, \tau), L_{x_1}^1 \subseteq L_{x_2}^2\} \to \mathcal{S}(\Omega_2^{L^2}, A, \alpha, \tau);$$

$$(x_2, (x_1, \rho \times V)) \mapsto (x_2, \mathrm{ind}_{L_{x_1}^1}^{L_{x_2}^2}(\rho \times V))$$

and

$$\mathrm{res}_{L_{x_1}^1}^{L_{x_2}^2} : \{(x_1, (x_2, \rho \times V)) \in \Omega_1 \times \mathcal{S}(\Omega_2^{L^2}, A, \alpha, \tau), L_{x_1}^1 \subseteq L_{x_2}^2\} \to \mathcal{S}(\Omega_1^{L^1}, A, \alpha, \tau);$$

$$(x_1, (x_2, \rho \times V)) \mapsto (x_2, \rho \times V|_{L_{x_1}^1})$$

are continuous

We are now going to introduce the notions of regularizations and of pairs of regular maps for a twisted covariant system.

DEFINITION 4.1.4. *Let (A, G, α, τ) be a twisted covariant system. A regularization of (A, G, α, τ) is a pair (Ω, R) consisting of a locally compact G-space Ω together with a G-equivariant continuous map $R : \widehat{A} \to \Omega$. If (Ω, R) is a regularization of (A, G, α, τ) and $L : \Omega \to \mathfrak{K}(G)$ is a continuous G-equivariant map, such that $N_\tau \subseteq L_x \subseteq S_x$ for all $x \in \Omega$, then the pair (R, L) is called a pair of regular maps with base Ω for (A, G, α, τ). Here the action of G on $\mathfrak{K}(G)$ is given by conjugation and S_x denotes the stabilizer of x for all $x \in \Omega$. If (R, L) is a pair of regular maps for (A, G, α, τ), then we define*

$$\mathcal{S}_R^L = \{(x, \rho \times V) \in C^*(\Omega^L, A, \alpha, \tau); \ker \rho \supseteq \ker R^{-1}(\{x\})\}$$

and

$$C^*(\mathcal{S}_R^L) = C^*(\Omega^L, A, \alpha, \tau)/I_{\mathcal{S}_R^L},$$

where $I_{\mathcal{S}_R^L} = \ker \mathcal{S}_R^L \subseteq C^*(\Omega^L, A, \alpha, \tau)$. $C^*(\mathcal{S}_R^L)$ is called the subgroup algebra defined by (R, L).

The definition of a regularization we use here is weaker than the definition given in [**13**, Definition 1], since there we also assumed that every primitive ideal was induced from a stability group of Ω (see [**13**, §2] for situations where both definitions coincide). The space $\mathcal{S}_R^L$ was first introduced in [**12**] in order to investigate the structure of the primitive ideal spaces of twisted covariant systems with continuously varying stabilizers. It is easily seen that

$$\mathcal{S}_R^L = \bigcup_{x \in \Omega} (A_x \rtimes_{\alpha, \tau} L_x)\widehat{\ }, \quad A_x = A/\ker R^{-1}(\{x\}).$$

The following proposition is a slight generalization of [**13**, Proposition 4].

PROPOSITION 4.1.5. *Suppose that (R, L) is a pair of regular maps with base Ω for the twisted system (A, G, α, τ). Then $\mathcal{S}_R^L$ is closed in $C^*(\Omega^L, A, \alpha, \tau)\widehat{\ }$.*

PROOF. Suppose that $((x_i, \rho_i \times V_i))_{i \in I}$ is a net in $\mathcal{S}_R^L$ which converges to some $(x, \rho \times V) \in C^*(\Omega^L, A, \alpha, \tau)\widehat{\ }$. Then $\rho_i \to \rho$ in $\mathrm{Rep}(A)$ by Proposition 4.1.3. Let $\pi \in \widehat{A}$ such that $\ker \pi \supseteq \ker \rho$. Then $\rho_i \to \pi$ by Proposition 1.2.1. It is then a consequence of Proposition 1.2.3 that, by passing over to a subnet if necessary, there exists a net $(\pi_i)_{i \in I} \subseteq \widehat{A}$ such that $\ker \pi_i \supseteq \ker \rho_i$ for each $i \in I$, and $\pi_i \to \pi$ in $\widehat{A}$. Since $\ker \pi_i \supseteq \ker \rho_i \supseteq \ker R^{-1}(\{x_i\})$ for each $i \in I$, it follows that $R(\pi_i) = x_i$ for all $i \in I$. Thus, by the continuity of R, we observe that $R(\pi) = x$. Since this is true for all $\pi \in \widehat{A}$ which are weakly contained in ρ, it follows that

$$\ker \rho = \bigcap \{\ker \pi; \pi \in \widehat{A} \text{ such that } \ker \pi \supseteq \ker \rho\} \supseteq \ker R^{-1}(\{x\}).$$

Hence $(x, \rho \times V) \in \mathcal{S}_R^L$. $\square$

As a consequence of Proposition 4.1.5 we observe that $\mathcal{S}_R^L$ is the dual space of $C^*(\mathcal{S}_R^L)$. We now want to define an action of G on $C^*(\mathcal{S}_R^L)$. Let $\tilde{G} = G/N_\tau$ and let $\delta : \Omega \times \tilde{G} \to \mathbb{R}^+$ be defined by the equation

$$\delta(x, \tilde{s}) = \left(\int_{\tilde{L}_{s^{-1}x}} g(\tilde{t}) \, d_{\tilde{L}_{s^{-1}x}} \tilde{t} \right) \cdot \left(\int_{\tilde{L}_x} g(\tilde{s}^{-1} \tilde{t} \tilde{s}) \, d_{\tilde{L}_x} \tilde{t} \right)^{-1}$$

for some $g \in C_c^+(\tilde{G})$ such that $g(\tilde{e}) \neq 0$. Note that in case where L_x is normal in G, $\delta(x, \tilde{s}) = \Delta_{\tilde{G}}(\tilde{s}) \Delta_{G/L_x}(\tilde{s}^{-1})$. We define an action, denoted γ^L, of G on $C^*(\Omega^L, A, \alpha, \tau)$ by

$$\gamma_s^L(f)(x, t) = \delta(x, \tilde{s}) \alpha_s(f(s^{-1}x, s^{-1}ts))$$

for all $f \in C_c(\Omega^L, A, \tau)$ (see [**12**, Section 4]). The corresponding action of G on $\mathcal{S}(\Omega^L, A, \alpha, \tau)$ is given by

$$(x, \rho \times V) \circ \gamma_{s^{-1}}^L = (sx, \rho \circ \alpha_{s^{-1}} \times V^s),$$

where $V_l^s = V_{s^{-1}ls}$ for all l in $L_{sx} = sL_x s^{-1}$. It is easily seen that $\mathcal{S}_R^L$ is a G-invariant subset of $C^*(\Omega^L, A, \alpha, \tau)$. Thus we obtain an action, also denoted γ^L, of G on $C^*(\mathcal{S}_R^L)$, which will be called the *canonical action of G on $C^*(\mathcal{S}_R^L)$.*

At this point it might be interesting to state the following property of $\mathcal{S}_R^L$ (see [**12**, Theorem 4]).

THEOREM 4.1.6. *Let (R, L) be a pair of regular maps for the twisted covariant system (A, G, α, τ). Then the map*

$$\text{Ind} : \mathcal{Q}_G(\mathcal{S}_R^L) \to \mathcal{I}(A \rtimes_{\alpha, \tau} G); \mathcal{Q}((x, \rho \times V)) \to \ker(\text{ind}_{L_x}^G (\rho \times V))$$

is a homeomorphism as a map onto its image.

We are now going to show that $C^*(\mathcal{S}_R^L)$ and the action γ^L of G on $C^*(\mathcal{S}_R^L)$ only depend on the composition $L \circ R : \widehat{A} \to \mathfrak{K}(G)$, but not on the particular maps R and L.

PROPOSITION 4.1.7. *Suppose that (R, L) and (R', L') are two pairs of regular maps for (A, G, α, τ) with bases Ω and Ω', respectively. If $L \circ R = L' \circ R'$, then $C^*(\mathcal{S}_R^L)$ is G-isomorphic to $C^*(\mathcal{S}_{R'}^{L'})$ with respect to the canonical actions of G.*

The proof of the following lemma goes back to some ideas used by Rieffel in the proof of [**63**, Proposition 8.1].

LEMMA 4.1.8. *Suppose that (Ω, R) is a regularization of (A, G, α, τ) such that Ω/G is Hausdorff. Then, for each $\pi \times U \in (A \rtimes_{\alpha, \tau} G)\widehat{\,}$, there exists a unique G-orbit $G(x) \in \Omega/G$ such that $\ker \pi \supseteq \ker R^{-1}(G(x))$.*

PROOF. Let $\pi \times U \in (A \rtimes_{\alpha, \tau} G)\widehat{\,}$ and let $\mathcal{F}$ denote the set of all G-invariant closed subsets F of Ω such that $\ker \pi \supseteq \ker R^{-1}(F)$. Then $\Omega \in \mathcal{F}$. Moreover, if $D = \bigcap_{F \in \mathcal{F}} F$ then $D \in \mathcal{F}$ (since $\ker \pi \supseteq \ker R^{-1}(F)$ for all $F \in \mathcal{F}$ implies that $\ker \pi \supseteq \ker R^{-1}(D)$).

We want to show that $D = G(x)$ for some $x \in \Omega$. Assume that D contains more than one G-orbit. Then it follows from the Hausdorff assumption for Ω/G that there exist two proper closed G-invariant subsets C_1 and C_2 of D such that $C_1 \cup C_2 = D$. It follows then from the construction of D that $\ker \pi \not\supseteq \ker R^{-1}(C_i)$ for $i = 1, 2$. But this contradicts $\ker \pi \supseteq \ker R^{-1}(D) = \ker R^{-1}(C_1) \cdot \ker R^{-1}(C_2)$. If y is another element of Ω such that $\ker \pi \supseteq \ker R^{-1}(G(y))$ then $\ker \pi \supseteq \ker R^{-1}(G(x) \cap G(y))$ from which follows that $G(x) = G(y)$. $\square$

Note that statements similar to Lemma 4.1.8 hold if Ω/G is almost Hausdorff or second countable (see [**13**, Proposition 3]).

PROOF OF PROPOSITION 4.1.7. Observe firstly that $C^*(\mathcal{S}_R^L)$ only depends on the closure, say $\tilde{\Omega}$, of $R(\widehat{A})$ in Ω. This follows easily from the fact that $\mathcal{S}_R^L$ is contained in $C^*(\tilde{\Omega}^L, A, \alpha, \tau)\widehat{\,}$, which in turn is a closed G-invariant subset of $C^*(\Omega^L, A, \alpha, \tau)\widehat{\,}$ by Proposition 4.1.2.

Suppose now that (R, L) and (R', L') are pairs of regular maps for (A, G, α, τ) such that $L \circ R = L' \circ R'$. We define $R' \times R : \widehat{A} \to \Omega' \times \Omega$ by $R' \times R(\rho) = (R'(\rho), R(\rho))$. Then it is clear that $(\Omega' \times \Omega, R' \times R)$ is also a regularization of (A, G, α, τ). The maps L and L' define canonical maps, also denoted L and

L', from $\Omega' \times \Omega \to \mathfrak{K}(G)$ by $(x, y) \mapsto L_y$ and $(x, y) \mapsto L'_x$, respectively, and it is clear that both pairs, $(R' \times R, L)$ and $(R' \times R, L')$ are pairs of regular maps for (A, G, α, τ) with base $\Omega' \times \Omega$. Since $L \circ R = L' \circ R'$ it follows easily that L and L' coincide on the closure of $R' \times R(\widehat{A})$ in $\Omega' \times \Omega$. Thus we conclude that $C^*(\mathcal{S}^L_{R' \times R})$ is G-isomorphic to $C^*(\mathcal{S}^{L'}_{R' \times R})$. Hence, by symmetry, the proof will be finished if we can show that $C^*(\mathcal{S}^L_R)$ is G-isomorphic to $C^*(\mathcal{S}^L_{R' \times R})$.

Consider the map

$$\Psi : C_c(\Omega') \odot C_c(\Omega^L, A, \tau) \to C_c((\Omega' \times \Omega)^L, A, \tau);$$

defined by

$$\Psi(f \otimes g)(x, y, s) = f(x)g(y, s).$$

It is easily seen that Ψ extends to a G-equivariant $*$-isomorphism between $C_0(\Omega') \otimes C^*(\Omega^L, A, \alpha, \tau)$ and $C^*((\Omega' \times \Omega)^L, A, \alpha, \tau)$ with respect to the diagonal action of G on $C_0(\Omega') \otimes C^*(\Omega^L, A, \alpha, \tau)$. Combining Ψ^{-1} with $\mathrm{id} \otimes \Phi$, where Φ denotes the quotient map from $C^*(\Omega^L, A, \alpha, \tau)$ onto $C^*(\mathcal{S}^L_R)$, we obtain a G-equivariant $*$-homomorphism from $C^*((\Omega' \times \Omega)^L, A, \alpha, \tau)$ onto $C_0(\Omega') \otimes C^*(\mathcal{S}^L_R)$.

We are now going to construct a G-equivariant continuous map, say Q, from $\mathcal{S}^L_R$ into Ω', which by Lemma 1.4.10 gives rise to a G-equivariant $*$-homomorphism Ψ_Q from $C_0(\Omega') \otimes C^*(\mathcal{S}^L_R)$ onto $C^*(\mathcal{S}^L_R)$. For this let $(y, \pi \times U) \in \mathcal{S}^L_R$. Then $(y, \pi \times U) \in (A_y \rtimes_{\alpha, \tau} L_y)\widehat{\ }$ with $A_y = A/\ker R^{-1}(\{y\})$. Let us restrict R' to $\widehat{A}_y = R^{-1}(\{y\})$, and let Λ be the closure of $R'(R^{-1}(\{y\}))$ in Ω'. Since $L'_x \subseteq S_x$ for all $x \in \Omega'$, and since $L' \circ R' = L \circ R$, we see that L_y acts trivially on Λ. Thus it follows from Lemma 4.1.8 that for each $\pi \times U \in (A_y \rtimes_{\alpha, \tau} L_y)\widehat{\ }$ there exists a unique element $x \in \Lambda \subseteq \Omega'$ such that $\ker \pi \supseteq \ker R'^{-1}(\{x\})$. We define $Q(y, \pi \times V) = x$ if and only if $\ker \pi \supseteq \ker R'^{-1}(\{x\})$, or, equivalently, $Q(y, \pi \times V) = R'(\rho)$ for any $\rho \in \widehat{A}$ with $\ker \rho \supseteq \ker \pi$.

It follows directly from the description of the G-action on $\mathcal{S}^L_R$ that Q is G-equivariant. In order to see that Q is continuous let $(y_i, \pi_i \times U_i) \to (y, \pi \times U)$ in $\mathcal{S}^L_R$. Then $\pi_i \to \pi$ in $\mathrm{Rep}(A)$, and $\pi_i \to \rho$ for any $\rho \in \widehat{A}$ with $\ker \rho \supseteq \ker \pi$ by Proposition 1.2.1. By passing to a subnet if necessary, we then find a net $(\rho_i)_{i \in I} \subseteq \widehat{A}$ such that $\ker \rho_i \supseteq \ker \pi_i$ for each $i \in I$ and $\rho_i \to \rho$ in $\widehat{A}$ (see Proposition 1.2.3). Thus $Q(\pi_i) = R'(\rho_i) \to R'(\rho) = Q(\pi)$.

Putting things together we see that there exists a G-equivariant $*$-homomorphism, say Φ, from $C^*((\Omega' \times \Omega)^L, A, \alpha, \tau)$ onto $C^*(\mathcal{S}^L_R)$, and the only thing which remains to show is that the kernel of Φ equals $I_{\mathcal{S}^L_{R' \times R}} = \ker \mathcal{S}^L_{R' \times R}$. For this let $((x, y), \pi \times U) \in C^*((\Omega' \times \Omega)^L, A, \alpha, \tau)\widehat{\ }$. It is clear that this representation corresponds to $(x, (y, \pi \times U)) \in \Omega' \times C^*(\Omega^L, A, \alpha, \tau)\widehat{\ } = (C_0(\Omega') \otimes C^*(\Omega^L, A, \alpha, \tau))\widehat{\ }$ via Ψ. By definition we have $((x, y), \pi \times U) \in \mathcal{S}^L_{R' \times R}$ if and only if $\ker \pi \supseteq \ker(R' \times R)^{-1}(\{(x, y)\}) = \ker(R'^{-1}(\{x\}) \cap R^{-1}(\{y\}))$. Thus, it follows that $((x, y), \pi \times U) \in \mathcal{S}^L_{R' \times R}$ if and only if $(y, \pi \times U) \in \mathcal{S}^L_R$ and $x = Q(y, \pi \times U)$, where Q is defined as above. On the other side we know from Lemma 1.4.10 that

$(y, \pi \times U) \circ \Psi_Q = (Q((y, \pi \times U), (y, \pi \times U)))$. Thus, $((x, y), \pi \times U) \in \mathcal{S}^L_{R' \times R}$ if and only if $(x, (y, \pi \times U)) \in \Psi_Q^*(\mathcal{S}^L_R)$, which easily implies that $\ker \Phi = \ker \mathcal{S}^L_{R' \times R}$. $\square$

4.2. Open regularizations and subgroup actions

Following the constructions of [**13**, Section 4] we want to give in this section an alternative construction of the subgroup algebras $C^*(\mathcal{S}^L_R)$ in the special situation where (R, L) is a pair of regular maps for the system (A, G, α, τ) with base Ω such that in addition $R : \widehat{A} \to \Omega$ is open and surjective. For this we have to recall the basic definitions of C^*-bundles.

Let Ω be a locally compact space. A *Banach bundle* $p : E \to \Omega$ consists of a continuous, open and surjective map from the topological Hausdorff space E onto Ω such that each *fiber* $A_x = p^{-1}(\{x\})$, $x \in \Omega$, is a Banach space and such that all Banach space operations on the fibers are continuous in E (for a more precise definition see [**25**, Section 1]). If all fibers are Banach $*$-algebras such that multiplication and involution on the fibers are continuous in E, then $p : E \to \Omega$ is called a *Banach $*$-algebra bundle*. If, in addition, all fibers A_x are C^*-algebras, then $p : E \to \Omega$ is called a C^*-bundle with base space Ω.

If $p : E \to \Omega$ is a C^*-bundle (resp. Banach $*$-algebra bundle), then the algebra $\Gamma_0(E)$ of all continuous sections $x \mapsto a(x) \in A_x$ such that $x \mapsto \|a(x)\|$ vanishes at infinity, equipped with pointwise multiplication and involution and the supremum norm, becomes a C^*-algebra (resp. Banach $*$-algebra). The irreducible representations of $A = \Gamma_0(E)$ are then given by the collection of all irreducible representations of the fibers A_x, where a representation ρ of A_x is viewed as a representation of A by defining $\rho(a) = \rho(a(x))$.

If $A = \Gamma_0(E)$ is the section algebra of a C^*-bundle $p : E \to \Omega$, then it was shown by Lee [**36**] that the resulting projection $P : \widehat{A} \to \Omega$ which maps $\rho \in \widehat{A}$ to the unique element $x \in \Omega$ such that $\rho \in \widehat{A}_x$ is always continuous, open and surjective. Conversely, it is also a result of Lee [**36**] that if A is any C^*-algebra such that there exists a continuous and open surjection from $\widehat{A}$ onto Ω, then there exists a C^*-bundle $p : E \to \Omega$ with fibers $A_x = A/\ker P^{-1}(\{x\})$ such that A is isomorphic to $\Gamma_0(E)$.

If $p : E \to \Omega$ is a C^*-bundle and $q : \Lambda \to \Omega$ is a continuous map, then the *bundle pull-back* q^*E of E by q is defined as the set $\{(y, e) \in \Lambda \times E; q(y) = p(e)\}$ together with the obvious projection from q^*E onto Λ. The continuous sections of q^*E may be identified with the continuous functions $f : \Lambda \to E$ such that $p(f(y)) = q(y)$ for all $y \in \Lambda$. If Λ is a locally closed (hence locally compact) subset of Ω, and $i : \Lambda \to \Omega$ denotes the inclusion map, then i^*E may be identified with $p^{-1}(\Lambda)$, and we will denote i^*E by $E|\Lambda$. Note that $\Gamma_0(E|\Lambda)$ is canonically isomorphic to the subquotient $\Gamma_0(E)_{P^{-1}(\Lambda)}$ defined by the locally closed subset $P^{-1}(\Lambda)$ of $\Gamma_0(E)\widehat{}$. The following definition of a twisted subgroup action was first given in [**13**] following some ideas of Raeburn and Williams given in [**56**]. Note that an (untwisted) subgroup action as defined below is just an action of the groupoid Ω^L on A as considered in [**61**].

DEFINITION 4.2.1. Let $p : E \to \Omega$ be a C^*-bundle, G a locally compact group and $L : \Omega \to \mathfrak{K}(G)$ a continuous map. A *subgroup action* of Ω^L on $A = \Gamma_0(E)$ is a family $\alpha = (\alpha^x)_{x \in \Omega}$ such that

(1) each α^x is a strongly continuous action of L_x on A_x,

(2) the map $\Omega^L \to E; (x, s) \mapsto \alpha^x_s(a(x))$ is continuous for each $a \in \Gamma_0(E)$.

If N_τ is a closed normal subgroup of G such that $N_\tau \subseteq L_x$ for all $x \in \Omega$, then a strictly continuous homomorphism $\tau : N_\tau \to \mathcal{U}(A)$ is called a *twisting map* for α, if each map $\tau^x : N_\tau \to \mathcal{U}(A_x)$ defined by

$$\tau^x_n a(x) = (\tau_n a)(x),$$

$a \in A$, $x \in \Omega$, is a twisting map for α^x. The pair $(\alpha, \tau) = (\alpha^x, \tau^x)_{x \in \Omega}$ is called a *twisted subgroup action* of Ω^L on A.

If (α, τ) is a twisted subgroup action of Ω^L on A, then we may define the *twisted subgroup crossed product* $A \rtimes_{\alpha,\tau} \Omega^L$ as follows: Let $\Gamma_c(q^*E, \tau)$ denote the set of all continuous sections f of the pull-back bundle q^*E, where $q : \Omega^L \to \Omega$ denotes projection to the first component, such that

$$f(x, ns) = f(x, s)\tau^x_{n^{-1}} \quad (x, s) \in \Omega^L, n \in N_\tau,$$

and such that $(x, \tilde{s}) \mapsto \|f(x, s)\|$ has compact support in $\Omega^{\tilde{L}}$, where, as usual, $\tilde{L}_x = L_x/N_\tau$ for all $x \in \Omega$. Similar to the construction of the subgroup algebra $C^*(\Omega^L, A, \alpha, \tau)$ we define multiplication, involution and norm on $\Gamma_0(q^*E, \tau)$ by

$$f * g(x, s) = \int_{\tilde{L}_x} f(x, r)\alpha^x_r(g(x, r^{-1}s)) \, d_{\tilde{L}_x}\tilde{r},$$

$$f^*(x, s) = \Delta_{\tilde{L}_x}(\tilde{s}^{-1})\alpha^x_s(f(x, s^{-1})^*)$$

and

$$\|f\|_1 = \sup_{x \in \Omega} \int_{\tilde{L}_x} \|f(x, s)\| \, d_{\tilde{L}_x}\tilde{s}.$$

It was shown in [**13**] (also using earlier results from [**56**]) that the completion of $\Gamma_c(q^*E, \tau)$ with respect to $\| \cdot \|_1$ is a Banach $*$-algebra, and we define $A \rtimes_{\alpha,\tau} \Omega^L$ as the enveloping C^*-algebra of this completion.

If $\rho \times V \in \text{Rep}(A_x \rtimes_{\alpha,\tau} L_x)$, then the pair $(x, \rho \times V)$ defines a representation of $A \rtimes_{\alpha,\tau} \Omega^L$ by

$$(x, \rho \times V)(f) = \pi \times U(f(x, \cdot)), \quad f \in \Gamma_c(q^*E, \tau).$$

The representations of $A \rtimes_{\alpha,\tau} \Omega^L$ obtained in this way are called the *subgroup representations* of $A \rtimes_{\alpha,\tau} \Omega^L$. The set of all equivalence classes of subgroup representations of $A \rtimes_\alpha \Omega^L$ (with dimension restricted by a fixed cardinal $\aleph$), equipped with the relative topology from $\text{Rep}(A \rtimes_{\alpha,\tau} \Omega^L)$ is denoted by $\mathcal{S}(A \rtimes_{\alpha,\tau} \Omega^L)$. Note that, as for $C^*(\Omega^L, A, \alpha, \tau)$, every irreducible representation of $A \rtimes_{\alpha,\tau} \Omega^L$ is a subgroup representation.

If $A \rtimes_\alpha \Omega^L$ is the non-twisted subgroup crossed product, then it is straight-forward to check that

$$\Phi : \Gamma_c(q^*E) \to \Gamma_c(q^*E, \tau); \quad \Phi f(x, s) = \int_{N_\tau} f(x, sn) \tau^x_{sns^{-1}} \, dn$$

extends to a $*$-homomorphism from $A \rtimes_\alpha \Omega^L$ onto $A \rtimes_{\alpha, \tau} \Omega^L$ such that the kernel of Φ is the intersection of all kernels of subgroup representations $(x, \rho \times V)$ of $A \rtimes_\alpha \Omega^L$ such that $\rho \times V$ preserves τ^x.

Note that if (A, G, α, τ) is a twisted covariant system and $L : \Omega \to \mathfrak{K}(G)$ is a continuous map such that $N_\tau \subseteq L_x$ for all $x \in \Omega$, then the subgroup algebra $C^*(\Omega^L, A, \alpha, \tau)$ is equal to the subgroup crossed product $C_0(\Omega, A) \rtimes_{\tilde{\alpha}, \tilde{\tau}} \Omega^L$, where $\tilde{\alpha}^x = \alpha$ and $\tilde{\tau}^x = \tau$ for all $x \in \Omega$. However, we found it convenient to do that special case first.

The most important example of a subgroup action comes from pairs of regular maps as in the following definition.

DEFINITION 4.2.2. Let (A, G, α, τ) be a twisted covariant system. A regularization (Ω, R) of (A, G, α, τ) is called an *open regularization* if $R : \widehat{A} \to \Omega$ is open and surjective. Moreover, a pair of regular maps (R, L) of (A, G, α, τ) with base Ω is called an *open pair of regular maps* if (Ω, R) is an open regularization of (A, G, α, τ).

If (R, L) is an open pair of regular maps for (A, G, α, τ) with base Ω, then, by Lee's theorem, we may write $A = \Gamma_0(E)$ for some C^*-bundle $r : E \to \Omega$ with fibers $A_x = A/\ker R^{-1}(\{x\})$. Since by the definition of regular pairs we have $N_\tau \subseteq L_x \subseteq S_x$, where S_x denotes the stabilizer of $x \in \Omega$, we see that that the restriction of α to L_x factors through an action, say α^x of L_x on A_x. If we denote the family $(\alpha^x)_{x \in \Omega}$ again by α, then it is easily seen (using [25, Proposition 10.3]) that (α, τ) becomes a twisted subgroup action of Ω^L on A (please compare with [13, Example 2]). We call this action *the restriction of* (α, τ) *to* Ω^L.

PROPOSITION 4.2.3. *If (R, L) is an open pair of regular maps for (A, G, α, τ) with base Ω, then $A \rtimes_{\alpha, \tau} \Omega^L$ is canonically isomorphic to $C^*(\mathcal{S}^L_R)$.*

PROOF. We assume first that τ is trivial. We define $\Phi : C_c(\Omega^L, A) \to \Gamma_c(q^*E)$ by $\Phi f(x, s) = f(x, s)(x)$. Straightforward computations show that Φ preserves multiplication and involution. If $g \in C_c(\Omega^L)$ and $a \in A$, then $\Phi(g \otimes a)(x, s) = g(x, s)a(x)$, and it is easily seen that these functions generate a dense subset of $\Gamma_c(q^*E)$ (see [56, Lemma 2.1]). If $\rho \times V$ is a representation of $A_x \rtimes_\alpha L_x$, i.e. $\ker \rho \supseteq \ker R^{-1}(\{x\})$, then it is a direct consequence of the definition of Φ that $(x, \rho \times V)(f) = (x, \rho \times V)(\Phi f)$, where on the left hand side we viewed $(x, \rho \times V)$ as a subgroup representation of $C^*(\Omega^L, A, \alpha)$ and on the right hand side we viewed $(x, \rho \times V)$ as a representation of $A \rtimes_\alpha \Omega^L$. Since $C^*(\mathcal{S}^L_R)$ is the quotient of $C^*(\Omega^L, A, \alpha)$ by the intersection of all those subgroup representations we conclude that the extension of Φ to $C^*(\Omega^L, A, \alpha, \tau)$ factors through an isomorphism from $C^*(\mathcal{S}_R^L)$ onto $A \rtimes_\alpha \Omega^L$. If τ is nontrivial, then the result follows by the fact that both twisted subgroup algebras are the quotients of the non-twisted algebras by the intersection of the kernels of all subgroup representations which preserve τ. $\square$

Note that in the remainder of this work it will be very convenient to work with both kinds of algebras, $C^*(\mathcal{S}_R^L)$ and $A \rtimes_{\alpha,\tau} \Omega^L$ defined by an open pair of regular maps (R, L). This is especially necessary when, in some situations, we start with an open pair of regular maps, so that we work with $A \rtimes_{\alpha,\tau} \Omega^L$, but where we then have to make a change of the base space to obtain a new pair of regular maps (R', L'), which is not necessarily open, but which satisfies the relation $L \circ R = L' \circ R'$ as in Proposition 4.1.7.

We close this section by recalling two propositions, which are both given in [13]. For the first one let (α, τ) be a twisted subgroup action of Ω^L on $A = \Gamma_0(E)$, and let Λ be a locally closed subset of Ω. Then we obtain a twisted subgroup action $(\alpha^\Lambda, \tau^\Lambda)$ of Λ^L on $A_{P^{-1}(\Lambda)} = \Gamma_0(E|\Lambda)$, by defining $(\alpha^\Lambda, \tau^\Lambda) = (\alpha^x, \tau^x)_{x \in \Lambda}$. The action $(\alpha^\Lambda, \tau^\Lambda)$ is called the *restriction* of (α, τ) to Λ. Again, we will denote $(\alpha^\Lambda, \tau^\Lambda)$ simply by (α, τ) if no confusion is possible. For the following result see [13, Proposition 8] (compare also with Proposition 4.1.2).

PROPOSITION 4.2.4. *Let (α, τ) be a twisted subgroup action of Ω^L on $A = \Gamma_0(E)$, Υ an open subset of Ω, and $\Lambda = \Omega \setminus \Upsilon$. Then inclusion of $\Gamma_c(q^*(E|\Upsilon), \tau)$ into $\Gamma_c(q^*E, \tau)$ and restriction of sections $f \in \Gamma_c(q^*E, \tau)$ to $q^*(E|\Lambda)$ extend to the subgroup crossed products and give rise to a short exact sequence*

$$0 \to A_{P^{-1}(\Upsilon)} \rtimes_{\alpha,\tau} \Upsilon^L \to A \rtimes_{\alpha,\tau} \Omega^L \to A_{P^{-1}(\Lambda)} \rtimes_{\alpha,\tau} \Lambda^L \to 0,$$

where $P : \widehat{A} \to \Omega$ denotes the canonical projection.

It is clear from the previous results that the canonical projection from $(A \rtimes_{\alpha,\tau} \Omega^L)\widehat{\,}$ onto Ω is always continuous. In case where $\tilde{L}_x = L_x/N_\tau$ is amenable for all $x \in \Omega$ we even get the following result which is [13, Corollary 4].

PROPOSITION 4.2.5. *Let (α, τ) be a twisted subgroup action of Ω^L on A such that $\tilde{L}_x = L_x/N_\tau$ is amenable for all $x \in \Omega$. Then the projection*

$$Q : (A \rtimes_{\alpha,\tau} \Omega^L)\widehat{\,} \to \Omega; (x, \pi \times U) \mapsto x$$

is continuous, open and surjective. Thus $A \rtimes_{\alpha,\tau} \Omega^L$ is isomorphic to the section algebra of a C^*-bundle $r : F \to \Omega$ with fibers equal to $A_x \rtimes_{\alpha^x, \tau^x} L_x$.

4.3. Decomposition of crossed products by subgroup algebras

In this section we want to show that one can decompose a twisted crossed product $A \rtimes_{\alpha,\tau} G$ by a pair (R, L) of regular maps with base Ω similar to the usual decomposition of a twisted crossed product by the decomposition action of G on $A \rtimes_{\alpha,\tau} M$ for a closed normal subgroup $M \supseteq N_\tau$ of G, at least if L_x is normal in G for all $x \in \Omega$. More precisely, we want to show that $A \rtimes_{\alpha,\tau} G$

is isomorphic to a quotient of $C^*(\mathcal{S}_R^L) \rtimes_{\gamma^L} G$, where γ^L denotes the canonical action of G on $C^*(\mathcal{S}_R^L)$ as defined in the previous section (see also [20, §2]).

DEFINITION 4.3.1. Suppose that Ω is a locally compact space, G is a locally compact group and $L : \Omega \to \mathfrak{K}(G); x \mapsto L_x$ is a continuous map. Then L defines an equivalence relation $\sim_L$ on $\Omega \times G$ by

$$(x, s) \sim_L (y, t) \Leftrightarrow x = y \text{ and } t \in sL_x.$$

The quotient space $\Omega \times_L G = (\Omega \times G)/\sim_L$ is called the *quotient of $\Omega \times G$ by* Ω^L.

Note that by [70, Lemma 2.3], $\Omega \times_L G$ is a locally compact Hausdorff space and the quotient map $\Omega \times G \to \Omega \times_L G$ is open. If $L : \Omega \to \mathfrak{K}(G)$ is constant, say $L_x = L$ for all $x \in \Omega$, then $\Omega \times_L G$ is equal to $\Omega \times G/L$. Thus we can understand $\Omega \times_L G$ as a generalization of taking quotients of groups by subgroups. In the following we will denote the $\sim_L$-equivalence class of $(x, s) \in \Omega \times G$ simply by $(x, \dot{s})$, where $\dot{s}$ stands for the left coset space sL_x.

Assume now that (R, L) is a pair of regular maps for (A, G, α, τ) with base Ω such that G acts trivially on Ω. We define a map $\mathsf{t} : \Omega^L \to \bigcup_{x \in \Omega} \mathcal{U}(A \rtimes_{\alpha,\tau} L_x)$ by $\mathsf{t}(x, l) = \tau_l^{L_x}$, where τ^{L_x} denotes the decomposition twisted map for L_x. Thus we have

$$(\mathsf{t}(x, l)f)(x, s) = \alpha_l(f(x, l^{-1}s)) \quad \text{and} \quad (f\mathsf{t}(x, l))(x, s) = f(x, sl^{-1})\Delta_{\tilde{G}}(l^{-1})$$

for all $l \in L_x$ and $f \in C_c(\Omega^L, A, \tau)$. We define $C_c(G, C_c(\Omega^L, A, \tau), \mathsf{t})$ as the set of all continuous A-valued functions f on $G \times \Omega^L$ satisfying
   (1) $f(s, \cdot, \cdot) \in C_c(\Omega^L, A, \tau)$ for all $s \in G$,
   (2) $f(hs, x, l) = f(s, x, lh)\Delta_{\tilde{G}}(h)$ for all $(s, x, l) \in G \times \Omega^L, n \in N_\tau$ and $h \in L_x$, and
   (3) the function $\Omega \times_L G \to \mathbb{R}^+; (x, \dot{s}) \mapsto \|f(s, x, \cdot)\|_1 = \int_{\tilde{L}_x} \|f(s, x, l)\| \, d_{\tilde{L}_x}\tilde{l}$
        has compact support.

Note that these conditions imply that $f(\cdot, x, \cdot) \in C_c(G, C_c(L_x, A, \tau), \tau^{L_x})$ for all $f \in C_c(G, C_c(\Omega^L, A, \tau), \mathsf{t})$. In order to define multiplication, involution and norm on $C_c(G, C_c(\Omega^L, A, \tau), \mathsf{t})$ we have to define a kind of smooth choice of Haar measures on the quotients of the closed normal subgroups of G. To this end let $(d_H)_{H \in \mathfrak{K}(G)}$ be a smooth choice of Haar measures on $\mathfrak{K}(G)$. For each normal $H \in \mathfrak{K}(G)$ let $d_{G/H}$ be the Haar measure on G/H which satisfies

$$\int_{G/H} \int_H g(sh) \, d_H h \, d_{G/H}\dot{s} = \int_G g(s) \, ds$$

for all $g \in C_c(G)$. The following lemma is a special case of [70, Lemma 2.22].

LEMMA 4.3.2. *Let $L : \Omega \to \mathfrak{K}(G)$ be a continuous map such that L_x is normal in G for all $x \in \Omega$ and let Haar measures on G/L_x be defined as above. For each $f \in C_c(\Omega \times_L G)$ define $\tilde{f}$ by $\tilde{f}(x) = \int_{G/L_x} f(x, \dot{s}) \, d_{G/L_x}\dot{s}$. Then $\tilde{f} \in C_c(\Omega)$.*

We want to use this lemma to define a norm on $C_c(G, C_c(\Omega^L, A, \tau), \mathfrak{t})$ by

$$\|f\|_1 = \sup_{x \in \Omega} \|f(\cdot, x, \cdot)\|_1^x,$$

where $\| \cdot \|_1^x$ denotes the norm on $C_c(G, C_c(L_x, A, \tau), \tau^{L_x})$ given by

$$\|g\|_1^x = \int_{G/L_x} \int_{\tilde{L}_x} \|g(s, l)\| \, d_{\tilde{L}_x} \tilde{l} d_{G/L_x} \dot{s}.$$

Note that $\| \cdot \|_1$ is well defined since by Lemma 4.3.2 the map $\Omega \to \mathbb{R}^+; x \mapsto \|f(\cdot, x, \cdot)\|_1^x$ is continuous with compact support. Let $L^1(G, L^1(\Omega^L, A, \tau), \mathfrak{t})$ denote the completion of $C_c(G, C_c(\Omega^L, A, \tau), \mathfrak{t})$ with respect to $\| \cdot \|_1$. By [25, Proposition 1.6] we conclude that $L^1(G, L^1(L_x, A, \tau), \mathfrak{t})$ is isomorphic to the Banach space $\Gamma_0(D)$ of all continuous sections which vanish at infinity of some Banach bundle $q : D \to \Omega$ with fibers $B_x = L^1(G, L^1(L_x, A, \tau), \tau^{L_x})$, the completion of $C_c(G, C_c(L_x, A, \tau), \tau^{L_x})$ with respect to $\| \cdot \|_1^x$. Let us note that $L^1(G, L^1(L_x, A, \tau), \tau^{L_x})$ is isometrically isomorphic to $L^1(G, A, \tau)$ via the extension of the canonical map

$$\Phi^x : C_c(G, A, \tau) \to C_c(G, C_c(L_x, A, \tau), \tau^{L_x});$$

$$\Phi^x(f)(s, l) = \Delta_{\tilde{G}}(s) \Delta_{G/L_x}(s^{-1}) f(ls)$$

(see also Example 1.1.1). On each fiber B_x we have a canonical multiplication and involution which makes B_x into a Banach $*$-algebra whose enveloping C^*-algebra, say $C^*(B_x)$, is isomorphic to $(A \rtimes_{\alpha, \tau} L_x) \rtimes_{\gamma^{L_x}, \tau^{L_x}} G$ and we know from Example 1.1.1 that each Φ^x extends to an isomorphism from $A \rtimes_{\alpha, \tau} G$ onto $C^*(B_x)$.

PROPOSITION 4.3.3. $L^1(G, L^1(\Omega^L, A, \tau), \mathfrak{t}) \cong \Gamma_0(D)$ becomes a Banach $*$-algebra if we define multiplication and involution fiberwise by the usual multiplication and involution on the fibers $B_x = L^1(G, L^1(L_x, A, \tau), \tau^{L_x})$. Moreover, its enveloping C^*-algebra, say $C^*(\Omega^L, A, \tau) \rtimes_{\gamma^L, \mathfrak{t}} G$, is canonically isomorphic to $C_0(\Omega, A \rtimes_{\alpha, \tau} G)$.

PROOF. We define

$$\Phi : C_c(\Omega) \odot C_c(G, A, \tau) \to C_c(G, C_c(\Omega^L, A, \tau), \mathfrak{t});$$

$$\Phi(f \otimes g)(s, x, l) = f(x) \Phi^x(g)(s, l),$$

where $\Phi^x : C_c(G, A, \tau) \to C_c(G, C_c(L_x, A, \tau), \tau^{L_x})$ is defined as above. Since the image of Φ^x is dense in B_x for each $x \in \Omega$, and since the image of Φ is invariant under pointwise multiplication with continuous functions on Ω, it follows from [25, Proposition 1.7] that Φ has dense image in $L^1(G, L^1(\Omega^L, A, \tau), \mathfrak{t})$. Moreover, since each Φ^x preserves multiplication and involution on the fibers, we see that fiberwise multiplication and involution is well defined on the image of Φ, and hence on all of $L^1(G, L^1(\Omega^L, A, \tau), \mathfrak{t})$. This implies that $L^1(G, L^1(\Omega^L, A, \tau), \mathfrak{t})$ is a Banach $*$-algebra. Since $L^1(G, L^1(\Omega^L, A, \tau), \mathfrak{t})$ is isomorphic to a section algebra

$\Gamma_0(D)$ of a Banach $*$-algebra bundle $q : D \to \Omega$ with fibers B_x we conclude that the irreducible representations of $L^1(G, L^1(\Omega^L, A, \tau), \mathfrak{t})$ are given by the collection of all irreducible representations of the fibers B_x in the canonical way. Since $C^*(B_x)$ is isomorphic to $(A \rtimes_{\alpha,\tau} L_x) \rtimes_{\gamma^{L_x}, \tau^{L_x}} G$ for all $x \in \Omega$, we may therefore parameterize the irreducible representations of $L^1(G, L^1(\Omega^L, A, \tau), \mathfrak{t})$ by the pairs $(x, (\rho \times U|_{L_x}) \times U)$ for $x \in \Omega$ and $\rho \times U \in (A \rtimes_{\alpha,\tau} G)\widehat{\ }$. Moreover, each Φ^x extends to the natural isomorphism between $A \rtimes_{\alpha,\tau} G$ and the iterated crossed product $(A \rtimes_{\alpha,\tau} L_x) \rtimes_{\gamma^{L_x}, \tau^{L_x}} G$, which implies easily that Φ extends to an isomorphism between $C_0(\Omega, A \rtimes_{\alpha,\tau} G)$ and $C^*(\Omega^L, A, \alpha, \tau) \rtimes_{\gamma^L, \mathfrak{t}} G$. $\square$

Assume now that γ^L is the canonical action of G on $C^*(\Omega^L, A, \alpha, \tau)$ as defined in the previous section (with respect to the trivial action of G on Ω — we still assume that L_x is normal in G for all $x \in \Omega$). It is easily seen that $(A \rtimes_{\alpha,\tau} L_x)\widehat{\ }$ is a G-invariant closed subset of $C^*(\Omega^L, A, \alpha, \tau)\widehat{\ }$, for each $x \in \Omega$, and that γ^L factors to the canonical action γ^{L_x} of G on the corresponding quotient $A \rtimes_{\alpha,\tau} L_x$ of $C^*(\Omega^L, A, \alpha, \tau)$. In particular, the canonical projection from $C^*(\Omega^L, A, \alpha, \tau)\widehat{\ }$ onto Ω becomes a regularization of the covariant system $(C^*(\Omega^L, A, \alpha, \tau), G, \gamma^L)$. Hence, Lemma 4.1.8 implies that each irreducible representation of $C^*(\Omega^L, A, \alpha, \tau) \rtimes_{\gamma^L} G$ factors through an irreducible representation of $(A \rtimes_{\alpha,\tau} L_x) \rtimes_{\gamma^{L_x}} G$ for some uniquely determined $x \in G$. Thus we may identify the set of irreducible representations of $C^*(\Omega^L, A, \alpha, \tau) \rtimes_{\gamma^L} G$ with the set of pairs $(x, \pi \times U)$ for some $\pi \times U \in ((A \rtimes_{\alpha,\tau} L_x) \rtimes_{\gamma^{L_x}} G)\widehat{\ }$. We say that a pair $(x, \pi \times U)$ *preserves* $\mathfrak{t}$ if and only if $\pi \times U$ preserves the decomposition twist τ^{L_x}. Recall from Example 1.1.1 that $\pi \times U$ preserves τ^{L_x} if and only if $\pi \times U$ is equivalent to a representation $(\rho \times V|_{L_x}) \times V$ for some representation $\rho \times V$ of $A \rtimes_{\alpha,\tau} G$.

PROPOSITION 4.3.4. *Let γ^L be as above. If $C^*(\Omega^L, A, \alpha, \tau) \rtimes_{\gamma^L, \mathfrak{t}} G$ is defined as in Proposition 4.3.3 then $C^*(\Omega^L, A, \alpha, \tau) \rtimes_{\gamma^L, \mathfrak{t}} G$ is canonically isomorphic to the quotient of $C^*(\Omega^L, A, \alpha, \tau) \rtimes_{\gamma^L} G$ by the intersection $I_{\mathfrak{t}}$ of all kernels of irreducible representations of $C^*(\Omega^L, A, \alpha, \tau) \rtimes_{\gamma^L} G$ which preserve $\mathfrak{t}$.*

PROOF. We define

$$\Psi : C_c(G, C_c(\Omega^L, A, \tau)) \to C_c(G, C_c(G, A, \tau), \mathfrak{t})$$

by

$$\Psi f(s, x, h) = \int_{L_x} f(sl, hsl^{-1}s^{-1}) \Delta_{\tilde{G}}(sl^{-1}s^{-1}) \, d_{L_x} l.$$

We observe that Ψ is given fiberwise by the natural maps

$$\Psi^x : C_c(G, C_c(L_x, A, \tau)) \to C_c(G, C_c(L_x, A, \tau), \tau^{L_x})$$

given by

$$\Psi^x g(s, l) = \int_{L_x} g(sl) \tau^{L_x}_{sls^{-1}} \, d_{L_x} l.$$

Since the image of Ψ^x is dense in $L^1(G, L^1(L_x, A, \tau), \tau^{L_x})$ for all $x \in \Omega$, and since the image of Ψ is invariant under multiplication with continuous functions on Ω, we conclude from [**25**, Proposition 1.7] that the image of Ψ is dense in $L^1(G, L^1(\Omega^L, A, \tau), \mathfrak{t})$ and hence in $C^*(\Omega^L, A, \alpha, \tau) \rtimes_{\gamma^L, \tau^L} G$. The proof follows now from the fact that each Ψ^x extends to a $*$-homomorphism from $(A \rtimes_{\alpha, \tau} L_x) \rtimes_{\gamma^{L_x}} G$ onto $(A \rtimes_{\alpha, \tau} L_x) \rtimes_{\gamma^{L_x}, \tau^{L_x}} G$ with kernel $I_{\tau^{L_x}} = \cap\{\ker(\pi \times U); \pi \times U \in (A \rtimes_{\alpha, \tau} L_x)\hat{}\}$. $\square$

We are now going back to our original problem.

PROPOSITION 4.3.5. *Let (R, L) be a pair of regular maps with base Ω for the twisted covariant system (A, G, α, τ) such that L_x is normal in G for all $x \in \Omega$. Then $A \rtimes_{\alpha, \tau} G$ is isomorphic to a quotient of $C^*(\mathcal{S}_R^L) \rtimes_{\gamma^L} G$.*

PROOF. First we reduce to the case where G acts trivially on Ω. For this let Ω' be the closure of $L(\Omega)$ in $\mathfrak{K}(G)$ and let $L' : \Omega' \to \mathfrak{K}(G)$ denote the inclusion. Further, let $R' = L \circ R$. Then (R', L') is a pair of regular maps for (A, G, α, τ) with base Ω' and with $L' \circ R' = L \circ R$. Since the normal subgroups of G are closed in $\mathfrak{K}(G)$, we see that G acts trivially on Ω'. By Proposition 4.1.7 we know that $C^*(\mathcal{S}_{R'}^{L'})$ is G-isomorphic to $C^*(\mathcal{S}_R^L)$. Thus by passing over to (R', L') we may assume that G acts trivially on Ω.

Let $C^*(\Omega^L, A, \alpha, \tau) \rtimes_{\gamma^L, \mathfrak{t}} G$ be as in Proposition 4.3.3. Then Proposition 4.3.3 implies that $C_0(\Omega, A \rtimes_{\alpha, \tau} G)$ is isomorphic to $C^*(\Omega^L, A, \alpha, \tau) \rtimes_{\gamma^L, \mathfrak{t}} G$, which in turn is a quotient of $C^*(\Omega^L, A, \alpha, \tau) \rtimes_{\gamma^L} G$ by Proposition 4.3.4. Looking at the definition of the homomorphism Ψ from $C^*(\Omega^L, A, \alpha, \tau) \rtimes_{\gamma^L} G$ onto $C^*(\Omega^L, A, \alpha, \tau) \rtimes_{\gamma^L, \mathfrak{t}} G$, as given in Proposition 4.3.4, and the isomorphism Φ from $C_0(\Omega, A \rtimes_{\alpha, \tau} G)$ onto $C^*(\Omega^L, A, \alpha, \tau) \rtimes_{\gamma^L, \mathfrak{t}} G$, as given in Proposition 4.3.3, we observe that a pair $(x, \pi \times U) \in C_0(\Omega, A \rtimes_{\alpha, \tau} G)\hat{}$ corresponds to the representation $(x, \pi \times U|_{L_x}) \times U$ of $C^*(\Omega^L, A, \alpha, \tau) \rtimes_{\gamma^L, \mathfrak{t}} G$, i.e., $(x, \pi \times U) \circ \Phi^{-1} \circ \Psi = (x, \pi \times U|_{L_x}) \times U$.

Since G acts trivially on Ω, it follows from Lemma 4.1.8 that for any $\pi \times U \in (A \rtimes_{\alpha, \tau} G)\hat{}$ there exists a unique element $x \in \Omega$ such that $\ker \pi \supseteq \ker R^{-1}(\{x\})$. Thus we may define a continuous map $Q : (A \rtimes_{\alpha, \tau} G)\hat{} \to \Omega$ by $Q(\pi \times U) = R(\rho)$ if and only if $\ker \rho \subseteq \ker \pi$. By Lemma 1.4.10 there exists a surjective $*$-homomorphism

$$\Psi_Q : C_0(\Omega, A \rtimes_{\alpha, \tau} G) \to A \rtimes_{\alpha, \tau} G$$

such that $\pi \times U \circ \Psi_Q = (x, \pi \times U)$ if $x = Q(\pi \times U)$. Thus we see that $\Theta = \Psi_Q \circ \Phi \circ \Psi$ becomes a surjective $*$-homomorphism from $C^*(\Omega^L, A, \alpha, \tau) \rtimes_{\gamma^L} G$ onto $A \rtimes_{\alpha, \tau} G$, and the proof will follow if we show that $I_{\mathcal{S}_R^L} \rtimes_{\gamma_L} G$ is contained in the kernel of Θ (note that $C^*(\mathcal{S}_R^L) \rtimes_{\gamma^L} G = (C^*(\Omega^L, A, \alpha, \tau) \rtimes_{\gamma^L} G)/(I_{\mathcal{S}_R^L} \rtimes_{\gamma_L} G))$.

But for $\pi \times U \in (A \rtimes_{\alpha, \tau} G)\hat{}$ we observe that

$$\pi \times U \circ \Theta = (x, \pi \times U|_{L_x}) \times U,$$

where $x = Q(\pi \times U)$. Thus we have to show that $\ker(x, \pi \times U|_{L_x}) \supseteq I_{\mathcal{S}_R^L}$ for all $\pi \times U \in (A \rtimes_{\alpha,\tau} G)^\frown$ and $x = Q(\pi \times U)$. This will follow if we can show that for any $(x, \rho \times V) \in C^*(\Omega^L, A, \alpha, \tau)^\frown$ with $\ker(x, \rho \times V) \supseteq \ker(x, \pi \times U|_{L_x})$ it follows that $(x, \rho \times V) \in \mathcal{S}_R^L$. But if $(x, \rho \times V)$ is as above, then it follows from the continuity of restriction that $\ker \rho \supseteq \ker \pi$, from which it follows that $x = Q(\pi \times U) = R(\rho)$ by the construction of Q. But this just means that $(x, \rho \times V) \in \mathcal{S}_R^L$. $\square$

4.4. Subgroup crossed products and duality

In this section we are going to investigate duality properties of subgroup crossed products which are related to abelian twisted systems. Suppose that (R, L) is a pair of regular maps for the abelian twisted system (A, G, α, τ) with base Ω. We say that (R, L) has a G-invariant base if G acts trivially on Ω. If this is the case, then it follows from Lemma 4.1.8 that there exists a continuous map, say $\widehat{R}$, from $(A \rtimes_{\alpha,\tau} G)^\frown$ into Ω, which is defined by

$$\widehat{R}(\pi \times U) = R(\rho) \Leftrightarrow \ker \rho \supseteq \ker \pi.$$

It follows directly from this definition that $\widehat{R}$ is $\widehat{\widetilde{G}}$-invariant. Furthermore, we define $L^\perp : \Omega \to \mathfrak{K}(\widehat{\widetilde{G}})$ by $L_x^\perp = (L_x)^\perp$. Since the map $\perp : \mathfrak{K}(\widetilde{G}) \to \mathfrak{K}(\widehat{\widetilde{G}}); H \mapsto H^\perp$ is continuous by [71] we see that $L^\perp$ is continuous, too. It follows that $(\widehat{R}, L^\perp)$ is a pair of regular maps for the dual system $(A \rtimes_{\alpha,\tau} G, \widehat{\widetilde{G}}, \widehat{\alpha})$ with $\widehat{\widetilde{G}}$-invariant base Ω.

DEFINITION 4.4.1. Let (R, L) be a pair of regular maps for the abelian twisted system (A, G, α, τ) with G-invariant base Ω, and let $(\widehat{R}, L^\perp)$ be as above. Then $(\widehat{R}, L^\perp)$ is said to be *the pair of regular maps for* $(A \rtimes_{\alpha,\tau} G, \widehat{\widetilde{G}}, \widehat{\alpha})$ *which is dual to* (R, L).

In what follows we want to relate $C^*(\mathcal{S}_R^L)$ with $C^*(\mathcal{S}_{\widehat{R}}^{L^\perp})$. Before we state our main result in this direction, let us define the dual action $\widehat{\alpha}$ of $\widehat{\widetilde{G}}$ on $C^*(\mathcal{S}_R^L)$ as follows: Suppose that (R, L) is a pair of regular maps for the twisted abelian system (A, G, α, τ) with base Ω. We define an action, also denoted $\widehat{\alpha}$, of $\widehat{\widetilde{G}}$ on $C^*(\Omega^L, A, \alpha, \tau)$ by

$$(\widehat{\alpha}_\chi f)(x, s) = \overline{\chi(s)} f(x, s)$$

for all $f \in C_c(\Omega^H, A, \tau)$. It is trivially seen that $\widehat{\alpha}$ extends to a strongly continuous action on $C^*(\Omega^L, A, \alpha, \tau)$, and it is also clear that $\mathcal{S}_R^L$ is a $\widehat{\widetilde{G}}$-invariant subset of $C^*(\Omega^L, A, \alpha, \tau)^\frown$. Thus $\widehat{\alpha}$ factors through an action of $\widehat{\widetilde{G}}$ on $C^*(\mathcal{S}_R^L)$.

The main result in this section is the following theorem.

THEOREM 4.4.2. *Suppose that* (R, L) *is a pair of regular maps for the twisted abelian system* (A, G, α, τ) *with G-invariant base Ω, and let $(\widehat{R}, L^\perp)$ be the pair of regular maps which is dual to (R, L). Then $C^*(\mathcal{S}_R^L)$ is Morita equivalent to $C^*(\mathcal{S}_{\widehat{R}}^{L^\perp})$ and the following is true*

(1) If $\widehat{\tilde{\alpha}}$ denotes the action of $\tilde{G}$ on $C^*(\mathcal{S}_{\frac{L^\perp}{R}})$ which is dual to the action $\widehat{\alpha}$ of G on $A \rtimes_{\alpha,\tau} G$, and if $q : G \to \tilde{G}$ denotes the quotient map, then $\widehat{\tilde{\alpha}} \circ q$ is Morita equivalent to the canonical action γ^L of G on $C^*(\mathcal{S}_R^L)$.

(2) The canonical action $\gamma^{L^\perp}$ of $\tilde{G}$ on $C^*(\mathcal{S}_{\frac{L^\perp}{R}})$ is Morita equivalent to the action $\widehat{\alpha}$ of $\tilde{G}$ on $C^*(\mathcal{S}_R^L)$.

The main tool for proving this theorem is the imprimitivity theorem for subgroup algebras as presented in [**12**, Theorem 2]. In order to recall this result in a way which fits to our needs, let us suppose that (A, G, α, τ) is a twisted covariant system and let $L : \Omega \to \mathcal{R}(G)$ be a continuous map such that $N_\tau \subseteq L_x$ for all $x \in \Omega$. Let $\Omega \times_L G$ denote the quotient of $\Omega \times G$ by Ω^L as defined in Definition 4.3.1. There is a canonical twisted action, say $(\tilde{\alpha}, \tilde{\tau})$ of G on $C_0(\Omega \times_L G, A)$, which is given by

$$\tilde{\alpha}_s(\varphi)(x, \dot{t}) = \alpha_s(\varphi(x, s^{-1}t)) \quad \text{and} \quad (\tilde{\tau}_n\varphi)(x, \dot{t}) = \tau_n(\varphi(x, \dot{t})).$$

Note that the maps

$$\Psi^x : C_c(G, C_c(\Omega \times_L G, A), \tilde{\tau}) \to C_c(G, C_c(G/L_x, A), \tilde{\tau}^x))$$

given by

$$\Phi^x(f)(s, \dot{t}) = f(s, x, \dot{t})$$

extend to surjective $*$-homomorphisms from $\mathcal{B} = C_0(\Omega \times_L G, A) \rtimes_{\tilde{\alpha},\tilde{\tau}} G$ onto the imprimitivity algebras $B^x = C_0(G/L_x, A) \rtimes_{\tilde{\alpha}^x, \tilde{\tau}^x} G$, where here $(\tilde{\alpha}^x, \tilde{\tau}^x)$ denotes the canonical twisted action of G on $C_0(G/L_x, A)$. Moreover, it follows from Lemma 4.1.8, that every irreducible representation of $C_0(\Omega \times_L G, A) \rtimes_{\tilde{\alpha}, \tilde{\tau}} G$ factors through an irreducible representation of B^x for some uniquely determined $x \in \Omega$.

In the following let $\mathcal{B}_0 = C_c(G, C_c(\Omega \times_L G, A), \tilde{\tau})$, viewed as a dense subalgebra of $\mathcal{B}$, let $\mathcal{X}_0 = C_c(\Omega \times G, A, \tau)$ and let $\mathcal{C}_0 = C_c(\Omega^L, A, \tau)$, the latter viewed as a dense subalgebra of $\mathcal{C} = C^*(\Omega^L, A, \alpha, \tau)$. We define left and right actions of $\mathcal{B}_0$ and $\mathcal{C}_0$ on $\mathcal{X}_0$ and $\mathcal{B}_0$- and $\mathcal{C}_0$- inner products on $\mathcal{X}_0$ by

$$
\begin{aligned}
F \cdot \xi(x, s) &= \int_{\tilde{G}} F(t, x, \dot{s}) \alpha_t(\xi(x, t^{-1}s)) \, d\tilde{t} \\
\xi \cdot f(x, s) &= \int_{\tilde{L}_x} \xi(x, sl) \alpha_{sl}(f(x, l^{-1})) \phi(x, l^{-1}) \, d_{\tilde{L}_x} \tilde{l} \\
\langle \xi, \eta \rangle_{\mathcal{B}_0}(s, x, \dot{r}) &= \int_{\tilde{L}_x} \xi(x, rl) \alpha_{rl}(\eta^*(x, l^{-1}r^{-1}s)) \, d_{\tilde{L}_x} \tilde{l} \\
\langle \xi, \eta \rangle_{\mathcal{C}_0}(x, l) &= \phi(x, l^{-1}) \int_{\tilde{G}} \xi^*(x, t) \alpha_t(\eta(x, t^{-1}l)) \, d\tilde{t},
\end{aligned}
$$

where $F \in \mathcal{B}_0$, $f \in \mathcal{C}_0$, $\xi, \eta \in \mathcal{X}_0$, and $\phi : \Omega^L \to \mathbb{R}^+$ is given by $\phi(x, l) = (\Delta_G(l)/\Delta_{L_x}(l))^{1/2}$. Note that the operations above are defined fiberwise by the operations which make $C_c(G, A, \tau)$ into a $B_0^x - C_0^x$ imprimitivity bimodule with $B_0^x = C_c(G, C_c(G/L_x, A), \tilde{\tau}^x)$ and $C_0^x = C_c(L_x, A, \tau)$.

PROPOSITION 4.4.3. *The above defined actions and inner products make $\mathcal{X}_0$ into a $\mathcal{B}_0 - \mathcal{C}_0$ pre-imprimitivity bimodule. Thus $\mathcal{X}_0$ may be completed to a $\mathcal{B} - \mathcal{C}$ imprimitivity bimodule $\mathcal{X}$.*

PROOF. If τ is trivial, then the Proposition is a direct consequence of [12, Theorem 2] together with [13, Lemma 1]. If τ is nontrivial, then the result follows from the fact that the canonical maps

$$\Phi^{\mathcal{B}} : C_c(G, C_c(\Omega \times_L G, A)) \to C_c(G, C_c(\Omega \times_L G, A), \tilde{\tau}),$$

$$\Phi^{\mathcal{C}} : C_c(\Omega^L, A) \to C_c(\Omega^L, A, \tau)$$

and

$$\Phi^{\mathcal{X}} : C_c(\Omega \times G, A) \to C_c(\Omega \times G, A, \tau),$$

all given by fiberwise integration against τ, transport all actions and inner products for the nontwisted systems to the actions and inner products as defined above (compare with Theorem 1.4.1). $\square$

Note that inducing a subgroup representation $(x, \rho \times V)$ of $C^*(\Omega^L, A, \alpha, \tau)$ to the representation $\mathrm{ind}^{\mathcal{X}}(x, \rho \times V)$ of $C_0(\Omega \times_L G, A) \rtimes_{\tilde{\alpha}, \tilde{\tau}} G$ via $\mathcal{X}$ is the same as taking the subgroup representation $(x, \mathrm{ind}^{X^x}(\rho \times V))$, where X^x denotes bimodule for inducing from L_x to G. (see [12, Proposition 5]). In what follows we want to show that in case where (A, G, α, τ) is an abelian twisted system, $C_0(\Omega \times_L G, A) \rtimes_{\tilde{\alpha}, \tilde{\tau}} G$ is isomorphic to the dual subgroup algebra $C^*(\Omega^{L^\perp}, A \rtimes_{\alpha, \tau} G, \hat{\alpha})$. We start with the following lemma (cf. [20, Lemma 3.3]).

LEMMA 4.4.4. *Suppose that G is an abelian locally compact group, Ω is a locally compact space, and $L : \Omega \to \mathfrak{K}(G)$ a continuous map. Let $L^\perp : \Omega \to \mathfrak{K}(\widehat{G})$ be defined as usual. Then*

$$\widehat{} : C_c(\Omega^L) \to C_0(\Omega \times_{L^\perp} \widehat{G}); f \mapsto \hat{f}$$

given by

$$\hat{f}(x, \mu) = \int_{L_x} f(x, s)\mu(s)\, d_{L_x}s$$

is well defined and extends to an isomorphism between $C^*(\Omega^L)$ and $C_0(\Omega \times_{L^\perp} \widehat{G})$.

PROOF. Since $\hat{f}(x, \mu) = (x, \mu)(f)$, for $f \in C_c(\Omega^L)$ and $(x, \mu) \in C^*(\Omega^L)\widehat{}$, the proof follows as soon as we have shown that the canonical map

$$p : \Omega \times \widehat{G} \to C^*(\Omega^L)\widehat{}; (x, \chi) \mapsto (x, \chi|_{L_x})$$

factors through a homeomorphism between $\Omega \times_{L^\perp} \widehat{G}$ and $C^*(\Omega^L)\widehat{}$. It is clear that p factors through a bijection between $\Omega \times_{L^\perp} \widehat{G}$ and $C^*(\Omega^L)\widehat{}$, and the continuity of p follows easily from the continuity of restricting representations. Thus it remains to show that p is open.

For this assume that $(x_i, \mu_i)_{i \in I}$ is a net in $C^*(\Omega^L)\widehat{}$ which converges to some $(x, \mu) \in C^*(\Omega^L)\widehat{}$, and let $\chi \in \widehat{G}$ such that $\chi|_{L_x} = \mu$. By continuity of induction (see Proposition 4.1.3) it follows that $\mathrm{ind}_{L_{x_i}}^G \mu_i \to \mathrm{ind}_{L_x}^G \mu$ in $\mathrm{Rep}(C^*(G))$, and by Proposition 1.4.9 we know that $\mathrm{ind}_{L_x}^G \mu = \lambda_{G/L_x} \otimes \chi$, where λ_{G/L_x} denotes the left regular representation of G/L_x. Since λ_{G/L_x} is weakly equivalent to $L_x^\perp$, we conclude that $\mathrm{ind}_{L_x}^G \mu$ is weakly equivalent to $\chi L_x^\perp$. The same argument shows that $\mathrm{ind}_{L_{x_i}}^G \mu_i$ is weakly equivalent to $\chi_i L_{x_i}^\perp$ for all $i \in I$, where the $\chi_i \in \widehat{G}$ are chosen such that $\chi_i|_{L_{x_i}} = \mu_i$. Using Propositions 1.2.1 and 1.2.3 we may pass to a subnet in order to assume that $\chi_i \nu_i \to \chi$ in $\widehat{G}$, for appropriately chosen characters $\nu_i \in L_{x_i}^\perp$. But this implies that $(x_i, \chi_i \nu_i) \to (x, \chi)$ in $\Omega \times \widehat{G}$. Since $p(x_i, \chi_i \nu_i) = (x_i, \mu_i)$, for all $i \in I$, this completes the proof. $\square$

PROPOSITION 4.4.5. *Let (A, G, α, τ) be an abelian twisted system and let $L : \Omega \to \mathfrak{K}(G)$ be a continuous map such that $N_\tau \subseteq L_x$ for all $x \in \Omega$ and let $L^\perp : \Omega \to \mathfrak{K}(\widehat{G})$ be defined as usual. Then the map*

$$\Phi : C_c(\Omega^{L^\perp}) \odot C_c(G, A, \tau) \to C_c(G, C_0(\Omega \times_L G, A), \tilde{\tau})$$

given by

$$\Phi(f \otimes g)(s, x, \dot{t}) = g(s)\widehat{f}(x, \dot{s}^{-1}\dot{t}),$$

where

$$\widehat{f}(x, \dot{s}) = \int_{L_x^\perp} f(x, \chi)\chi(\dot{s}) \, d_{L_x^\perp} \chi, \quad f \in C_c(\Omega^{L^\perp}),$$

with respect to a smooth choice of Haar measures on $\mathfrak{K}(\widehat{G})$, extends to an isomorphism between $C^*(\Omega^{L^\perp}, A \rtimes_{\alpha, \tau} G, \widehat{\alpha})$ and $C_0(\Omega \times_L G, A) \rtimes_{\tilde{\alpha}, \tilde{\tau}} G$.

PROOF. It follows from Lemma 4.4.4 applied to the map $L^\perp : \Omega \to \mathfrak{K}(\widehat{G})$ that Φ is well defined. By Proposition 4.2.5 we know that $C^*(\Omega^{L^\perp}, A \rtimes_{\alpha, \tau} G, \widehat{\alpha})$ is isomorphic to $\Gamma_0(D)$ for some C^*-bundle $r : D \to \Omega$ with fibers $(A \rtimes_{\alpha, \tau} G) \rtimes_{\widehat{\alpha}} L_x^\perp$, and it is also a consequence of Proposition 4.2.5, using also Proposition 4.4.3, that $C_0(\Omega \times_L G, A) \rtimes_{\tilde{\alpha}, \tilde{\tau}} G$ is isomorphic to $\Gamma_0(E)$ for some C^*-bundle $p : E \to \Omega$ with fibers $C_0(G/L_x, A) \rtimes_{\tilde{\alpha}^x, \tilde{\tau}^x} G$. If $f \otimes g \in C_c(\Omega^{L^\perp}) \odot C_c(G, A, \tau)$, then the corresponding section is given by $x \mapsto f^x \otimes g$, where $f^x(\chi) = f(x, \chi)$, while for for any $h \in C_c(G, C_0(\Omega \times_L G, A), \tilde{\tau})$ the corresponding section in $\Gamma_0(E)$ is given by $x \mapsto h^x = h(\cdot, x, \cdot)$.

For each $x \in \Omega$ let

$$\Phi^x : (A \rtimes_{\alpha, \tau} G) \rtimes_{\widehat{\alpha}} L_x^\perp \to C_0(G/L_x, A) \rtimes_{\tilde{\alpha}^x, \tilde{\tau}^x} G$$

denote the canonical isomorphism as given in [**16**, Proposition 2.2]. Then it follows directly from the definitions that $\Phi(f \otimes g)^x = \Phi^x(f^x \otimes g)$. Since each Φ^x is an isomorphism, and since the image of Φ is invariant under pointwise multiplication with functions in $C_0(\Omega)$ we easily conclude that Φ extends to an isomorphism between $C^*(\Omega^{L^\perp}, A \rtimes_{\alpha, \tau} G, \widehat{\alpha})$ and $C_0(\Omega \times_L G, A) \rtimes_{\tilde{\alpha}, \tilde{\tau}} G$. $\square$

We are now ready to prove the main result of this section.

PROOF OF THEOREM 4.4.2. Let (R, L) be a pair of regular maps for the abelian twisted system (A, G, α, τ) with G-invariant base Ω, and let $(\widehat{R}, L^\perp)$ be the regular pair for $(A \rtimes_{\alpha,\tau} G, \widehat{G}, \widehat{\alpha})$ which is dual to (R, L). By Proposition 4.4.5, together with Proposition 4.4.3, we know that $\mathcal{X}_0 = C_c(\Omega \times G, A, \tau)$ can be completed to a $C^*(\Omega^{L^\perp}, A \rtimes_{\alpha,\tau} G, \widehat{\alpha}) - C^*(\Omega^L, A, \alpha, \tau)$ imprimitivity bimodule $\mathcal{X}$, such that the actions and inner products are given on each fiber $X_0^x = C_c(G, A, \tau)$ by the actions and inner products which make (the completion of) $C_c(G, A, \tau)$ into a $(A \rtimes_{\alpha,\tau} G) \rtimes_{\widehat{\alpha}} L_x^\perp - A \rtimes_{\alpha,\tau} L_x$ imprimitivity bimodule (compare with the proof of [16, Proposition 2.1]).

We define actions u and $\widehat{u}$ of G on X_0 by

$$u_s(\xi)(x, t) = \Delta_G(s) \Delta_{G/L_x}(s) \xi(x, ts)$$

and

$$\widehat{u}_\chi(\xi)(x, s) = \overline{\chi(s)} \xi(x, s).$$

Then u is defined fiberwise by the actions, say u^x, of G on the fibers $C_c(G, A, \tau)$ which implement the Morita equivalences between the canonical actions γ^{L_x} of G on $A \rtimes_{\alpha,\tau} L_x$ and $\widehat{\widehat{\alpha^x}} \circ q^x$ of G on $(A \rtimes_{\alpha,\tau} G) \rtimes_{\widehat{\alpha}} L_x^\perp$, where $q^{L_x}: G \to G/L_x$ denotes the quotient map (see Proposition 2.2.1). Similarly, $\widehat{u}$ is defined fiberwise by the actions, say $\widehat{u}^x$, which implement Morita equivalences between the decomposition action $\gamma^{L_x^\perp}$ of $\widehat{G}$ on $(A \rtimes_{\alpha,\tau} G) \rtimes_{\widehat{\alpha}} L_x^\perp$ and the action $\widehat{\alpha|_{L_x}} \circ q^{L_x^\perp}$. Since the actions of G and $\widehat{G}$ on $C^*(\Omega^L, A, \alpha, \tau)$ and $C^*(\Omega^{L^\perp}, A \rtimes_{\alpha,\tau} G, \widehat{\alpha})$, respectively, are also defined fiberwise, we conclude that γ^L is Morita equivalent to $\widehat{\widehat{\alpha}} \circ q$ and that $\gamma^{L^\perp}$ is Morita equivalent to $\widehat{\alpha}$.

Since the corresponding actions of G and $\widehat{G}$ on $C^*(\mathcal{S}_R^L)$ and $C^*(\mathcal{S}_{\widehat{R}}^{L^\perp})$ are the images of the actions of G and $\widehat{G}$ on $C^*(\Omega^L, A, \alpha, \tau)$ and $C^*(\Omega^{L^\perp}, A \rtimes_{\alpha,\tau} G, \widehat{\alpha})$ under the quotient maps, respectively, the theorem is proved as soon as we have shown that a representation $(x, \rho \times V) \in C^*(\Omega^L, A, \alpha, \tau)\widehat{}$ is an element of $\mathcal{S}_R^L$ if and only if $\mathrm{ind}^{\mathcal{X}}(x, \rho \times V) \in \mathcal{S}_{\widehat{R}}^{L^\perp}$. But since induction via $\mathcal{X}$ is the same as inducing $\rho \times V$ to $C_0(G/L_x, A) \rtimes_{\widetilde{\alpha}^x, \widetilde{\tau}^x} G$ via the usual inducing process, and then transporting this representation to $(A \rtimes_{\alpha,\tau} G) \rtimes_{\widehat{\alpha}} L_x^\perp$ via the canonical isomorphism Φ^x, we conclude that $\mathrm{ind}^{\mathcal{X}}(x, \rho \times V) = (x, \mathrm{ind}_{L_x}^G(\rho \times V) \times W)$ for some unitary representation W of $L_x^\perp$, where $\mathrm{ind}_{L_x}^G(\rho \times V)$ denotes the usual induced representation of $A \rtimes_{\alpha,\tau} G$ (see the constructions preceding [16, Theorem 2.3]). It follows that $\mathrm{ind}^{\mathcal{X}}(x, \rho \times V) \in \mathcal{S}_{\widehat{R}}^{L^\perp}$ if and only if $\ker(\mathrm{ind}_{L_x}^G(\rho \times V)) \supseteq \ker \widehat{R}(\{x\})$. But this is equivalent to saying that $\widehat{R}(\pi \times U) = x$ for all $\pi \times U \in (A \rtimes_{\alpha,\tau} G)\widehat{}$ such that $\ker(\pi \times U) \supseteq \ker(\mathrm{ind}_{L_x}^G(\rho \times V))$. But since $\ker(\mathrm{res}_{N_x}^G(\mathrm{ind}_{L_x}^G(\rho \times V))) = \bigcap_{s \in G} \ker \rho \circ \alpha_s$ by Proposition 1.4.8, it follows from the continuity of restricting representations, that $\ker \pi \supseteq \bigcap_{s \in G} \ker \rho \circ \alpha_s$ for all such π. Thus, if $\ker \rho \supseteq \ker R^{-1}(\{x\})$, then it follows easily from the definition of $\widehat{R}$ and the continuity of R that $\widehat{R}(\pi \times U) = x$ for all such $\pi \times U$. On the other side, since $\ker(\mathrm{ind}^G_{L_x}(\rho \times V))$ is equal to the intersection of all kernels of irreducible representations $\pi \times U$ such $\ker(\pi \times U) \supseteq \ker(\mathrm{ind}^G_{L_x}(\rho \times V))$ we conclude that

$$\bigcap_{s \in G} \ker \rho \circ \alpha_s = \cap\{\ker \pi; \ker(\pi \times U) \supseteq \ker(\mathrm{ind}^G_{L_x}(\rho \times V))\}.$$

If $\widehat{R}(\pi \times U) = x$ for all $\pi \times U$ which are weakly contained in $\mathrm{ind}^G_{L_x}(\rho \times V)$, it follows from this and the definition of $\widehat{R}$, that $R(\sigma) = x$ for all $\sigma \in \widehat{A}$ such that $\ker \sigma \supseteq \bigcap_{s \in G} \ker \rho \circ \alpha_s$. In particular, we see that $R(\sigma) = x$ for all $\sigma \in \widehat{A}$ such that $\ker \sigma \supseteq \ker \rho$, which implies that $\ker \rho \supseteq \ker R^{-1}(\{x\})$, and hence $(x, \rho \times V) \in \mathcal{S}^L_R$. $\square$

We want to close this section with a result which allows to pass over to Morita equivalent twisted actions while investigating subgroup algebras. For this let (α, τ) and (β, σ) be twisted actions of G on the C^*-algebras A and B, respectively, and suppose that (X, u) is a $(\beta, \sigma) - (\alpha, \tau)$ imprimitivity pair. Assume further that (R, L) is a pair of regular maps for (A, G, α, τ) with base Ω. We define $R' : \widehat{B} \to \Omega$ by $R'(\mathrm{ind}^{(X,u)} \rho) = R(\rho)$. Then (R', L) is a pair of regular maps for (B, G, β, σ). Let $\gamma^{L,\alpha}$ and $\gamma^{L,\beta}$ denote the canonical actions of G on $C^*(\mathcal{S}^L_R)$ and $C^*(\mathcal{S}^L_{R'})$, respectively. Then we have

PROPOSITION 4.4.6. *Let $\gamma^{L,\alpha}$ and $\gamma^{L,\beta}$ be as above. Then $\gamma^{L,\alpha}$ and $\gamma^{L,\beta}$ are Morita equivalent. Furthermore, if $\tilde{G} = G/N_\tau$ is abelian, and $\widehat{\alpha}$ and $\widehat{\beta}$ denote the dual actions of $\widehat{\tilde{G}}$ on $C^*(\mathcal{S}^L_R)$ and $C^*(\mathcal{S}^L_{R'})$, respectively, then $\widehat{\alpha}$ is Morita equivalent to $\widehat{\beta}$.*

PROOF. Let $\mathcal{B}_0 = C_c(\Omega^L, B)$, $\mathcal{A}_0 = C_c(\Omega^L, A)$ and $\mathcal{X}_0 = C_c(\Omega^L, X)$. We define left and right actions of $\mathcal{B}_0$ and $\mathcal{A}_0$ on $\mathcal{X}_0$, and $\mathcal{B}_0$- and $\mathcal{A}_0$-valued inner products on $\mathcal{X}_0$ by

$$f \cdot \xi(x, s) = \int_{L_x} f(x, t) u_t(\xi(x, t^{-1}s)) \, d_{L_x} t$$

$$\xi \cdot g(x, s) = \int_{L_x} \xi(x, t) \alpha_t(g(x, t^{-1}s)) \, d_{L_x} t$$

$$\langle \xi, \eta \rangle_{\mathcal{A}_0}(x, s) = \int_{L_x} \alpha_{t^{-1}}(\langle \xi(x, t), \eta(x, ts) \rangle_A) \, d_{L_x} t$$

$$\langle \xi, \eta \rangle_{\mathcal{B}_0}(x, s) = \int_{L_x} \Delta_{L_x}(s^{-1}t) \langle \xi(x, t), u_t(\eta(x, s^{-1}t)) \rangle_B \, d_{L_x} t,$$

where $f \in \mathcal{B}_0$, $g \in \mathcal{A}_0$ and $\xi, \eta \in \mathcal{X}_0$. The operations above are given fiberwise by the operations which make $C_c(L_x, X)$ into a $C_c(L_x, B) - C_c(L_x, A)$ imprimitivity bimodule (see Proposition 2.1.8). It is straightforward to check that $\mathcal{X}_0$ satisfies all necessary conditions for extending the actions and inner products to the completion, say $\mathcal{X}$, of $\mathcal{X}_0$ such that $\mathcal{X}$ becomes a $C^*(\Omega^L, B, \beta) - C^*(\Omega^L, A, \alpha)$

imprimitivity bimodule, and it is easily seen that inducing a subgroup representation $(x, \rho \times V)$ of $C^*(\Omega^L, A, \alpha)$ to $C^*(\Omega^L, B, \beta)$ via $\mathcal{X}$ is the same as taking the subgroup representation $(x, \text{ind}^{X \rtimes_u L_x}(\rho \times V))$, where $X \rtimes_u L_x$ denotes the canonical $B \rtimes_\beta L_x - A \rtimes_\alpha L_x$ imprimitivity bimodule.

We denote the canonical actions of G on $C^*(\Omega^L, A, \alpha)$ and $C^*(\Omega^L, B, \beta)$ also by $\gamma^{L,\alpha}$ and $\gamma^{L,\beta}$, respectively. We define an action v of G on $\mathcal{X}$ by

$$v_s(\xi)(x, t) = \delta(x, s) u_s(\xi(s^{-1}x, s^{-1}ts))$$

for all $s \in G$ and $\xi \in \mathcal{X}_0$. Exactly as in the proof of Proposition 2.1.8 we check that

$$(v_s(\xi)\langle v_s(\eta), v_s(\zeta)\rangle_{\mathcal{A}_0})(x, t) = v_s(\xi\langle\eta, \zeta\rangle_{\mathcal{A}_0})(x, t)$$

for all $s \in G$ and $\xi, \eta, \zeta \in \mathcal{X}_0$. Thus $v_s \in \text{Aut}(\mathcal{X})$ for all $s \in G$. It is also easily seen that

$$\langle v_s(\xi), v_s(\eta)\rangle_{\mathcal{A}_0} = \gamma_s^{L,\alpha}(\langle\xi, \eta\rangle_{\mathcal{A}_0})$$

and

$$\langle v_s(\xi), v_s(\eta)\rangle_{\mathcal{B}_0} = \gamma_s^{L,\beta}(\langle\xi, \eta\rangle_{\mathcal{B}_0})$$

for all $s \in G$ and $\xi, \eta \in \mathcal{X}_0$. Thus $(\mathcal{X}, v)$ is a $\gamma^{L,\beta} - \gamma^{L,\alpha}$ imprimitivity pair. Since $\text{ind}^{\mathcal{X}}(x, \rho \times V)$ is equivalent to $(x, \text{ind}^{X \rtimes_u L_x}(\rho \times V))$, for all subgroup representations $(x, \rho \times V)$ of $C^*(\Omega^L, A, \alpha)$, we see that $(x, \rho \times V)$ preserves τ if and only if $\text{ind}^{\mathcal{X}}(x, \rho \times V)$ preserves σ. Thus we may pass to a suitable quotient of $\mathcal{X}$ to obtain a $C^*(\Omega^L, B, \beta, \sigma) - C^*(\Omega^L, A, \alpha, \tau)$ imprimitivity bimodule, say $\mathcal{X}^\tau$, and an action, also denoted v, of G on $\mathcal{X}^\tau$ such that $(\mathcal{X}^\tau, v)$ becomes a $\gamma^{L,\beta} - \gamma^{L,\alpha}$ imprimitivity pair, where we view $\gamma^{L,\beta}$ and $\gamma^{L,\alpha}$ as actions on the twisted subgroup algebras $C^*(\Omega^L, A, \beta, \sigma)$ and $C^*(\Omega^L, A, \alpha, \tau)$, respectively (see Proposition 2.1.2). Finally, it follows directly from the definition of the map $R' : \widehat{B} \to \Omega$, and the fact that induction via $X \rtimes_u L_x$ is the same as induction via the imprimitivity pair $(x, u|_{L_x})$, that inducing via $\mathcal{X}^\tau$ maps $\mathcal{S}_R^L$ onto $\mathcal{S}_{R'}^L$. Thus, again by passing over to a suitable quotient, say $\mathcal{Y}$, we conclude that $\gamma^{L,\alpha}$ and $\gamma^{L,\beta}$ are Morita equivalent actions of G on $C^*(\mathcal{S}_R^L)$ and $C^*(\mathcal{S}_{R'}^L)$, respectively.

In order to complete the proof, assume that $\tilde{G} = G/N_\tau$ is abelian and let $\mathcal{X}$ be as above. Then we define an action, say $\widehat{u}$, of $\widehat{\tilde{G}}$ on $\mathcal{X}_0$ by

$$\widehat{u}_\chi(\xi)(x, s) = \overline{\chi(s)}\xi(x, s),$$

$\chi \in \widehat{\tilde{G}}$, $s \in G$ and $\xi \in \mathcal{X}_0$. Similar, but easier arguments as used above show that $\widehat{u}$ extends to an action on $\mathcal{X}$ which factors through an action of $\widehat{\tilde{G}}$ on the $C^*(\mathcal{S}_{R'}^L) - C^*(\mathcal{S}_R^L)$ imprimitivity bimodule $\mathcal{Y}$ as constructed above, implementing a Morita equivalence between $\widehat{\beta}$ and $\widehat{\alpha}$. $\square$

CHAPTER 5

Crossed products with continuous trace

Let (A, G, α, τ) be a separable twisted abelian system such that A has continuous trace. In this chapter we are going to prove necessary and sufficient conditions for $A \rtimes_{\alpha,\tau} G$ to have continuous trace. In the first section we start with systems (A, G, α, τ) such that $\widehat{A}$ is a σ-proper G-space and show that under this assumption $A \rtimes_{\alpha,\tau} G$ has continuous trace whenever the twisted subgroup crossed product $A \rtimes_{\alpha,\tau} \Omega^S$ has continuous trace. The second section is devoted to the study of subgroup crossed products by pointwise unitary subgroup actions. We then use these results in the following sections in order to prove the theorems as stated in the introduction.

5.1. Subgroup crossed products and σ-proper G-spaces

Suppose that Ω is a locally compact space, G is a locally compact group, and $H : \Omega \to \mathfrak{K}(G)$ is a continuous map. Recall from Definition 4.3.1 that the quotient $\Omega \times_H G$ of $\Omega \times G$ by Ω^H consists of all pairs $(x, \dot{s})$, where $\dot{s} = sH_x$ denotes the left H_x-coset of s.

DEFINITION 5.1.1. Suppose that (G, Ω) is a locally compact transformation group. Then the action of G on Ω is called σ-proper, if the following conditions are satisfied:
   (1) The stabilizer map $S : \Omega \to \mathfrak{K}(G); x \mapsto S_x$ is continuous.
   (2) The map $p : \Omega \times_S G \to \Omega \times \Omega; p(x, \dot{s}) = (x, sx)$ is proper, i.e. $p^{-1}(C)$ is compact for any compact set $C \subseteq \Omega \times \Omega$.

If Ω is a σ-proper G-space, then we say that Ω is σ-trivial, if there exists a continuous section $\mathfrak{s} : \Omega/G \to \Omega$. Moreover, a σ-proper G-space Ω is called locally σ-trivial if Ω has local sections. This means that each $\omega \in \Omega$ has a neighborhood U of $q(\omega)$ in Ω/G such that there exists a continuous section $\mathfrak{s}_U : U \to q^{-1}(U)$, where $q : \Omega \to \Omega/G$ denotes the quotient map.

The reason for using "σ" in the notation of the definition above is that in [56], where the notion of σ-properness was defined first, σ always denoted the stabilizer map from $\Omega \to \mathfrak{K}(G)$. Since we call the stabilizer map usually (but not always) by the letter S, a direct translation of the terminology of [56] would result in the notation of S-properness instead of σ-properness. However, since we feel no need to specify a name for the stabilizer map in our notation, we use the letter σ just to distinguish from usual proper G-spaces. This terminology also agrees with the terminology used in [13, 14, 20].

If G acts σ-properly on Ω, then Ω/G is Hausdorff. This follows from the arguments used in the proof of [56, Proposition 4.3, (2)]. If the action of G on Ω is free, i.e., if all stabilizers are trivial, then G acts σ-properly on Ω if and only if the action of G on Ω is proper in the usual sense. This follows directly from the definitions. More generally, if S_x is equal to a fixed subgroup, say N, of G, then saying that the action of G on Ω is σ-proper is equivalent to saying that G/N acts properly on Ω.

Our definition of σ-triviallity of a G-space is not exactly the same as that given in [56, Definition 4.2]. The reason is that the definition given in [56] only makes sense for abelian groups G, while our definition works in general. However, in case where G is abelian both definitions coincide by [56, Proposition 4.3]. The following proposition also follows from arguments used in [56, Proposition 4.3].

PROPOSITION 5.1.2. *Suppose that $H : \Omega \to \mathfrak{K}(G)$ is a continuous map. Let G act on $\Omega \times_H G$ by left translation on the second variable. Then $\Omega \times_H G$ is a σ-trivial G-space. Conversely, if Ω is any σ-trivial G-space with continuous section $\mathfrak{s} : \Omega/G \to \Omega$, then Ω is G-homeomorphic to $\Omega/G \times_{S \circ \mathfrak{s}} G$, where $S : \Omega \to \mathfrak{K}(G)$ denotes the stabilizer map.*

One fundamental result for proper actions is Palais's slice theorem which says that, if G is a Lie group, then every separable proper G-space is locally trivial. Unfortunately, there is no reasonable class of groups G such that a similar result holds also for σ-proper G-spaces. This follows from [56, Example 4.7], where Raeburn and Williams constructed a σ-proper action of $\mathbb{R}$ on $\mathbb{C}^2$ which is not locally σ-trivial. However, we see later that the following very weak version of the slice theorem is sufficient for our purposes. It shows that every σ-proper G-action is σ-trivial on "closures of subsequences".

PROPOSITION 5.1.3. *Suppose that G is a separable locally compact group and that Ω is a second countable locally compact G-space such that the stabilizer map $S : \Omega \to \mathfrak{K}(G); x \mapsto S_x$ is continuous. Let $q : \Omega \to \Omega/G$ denote the quotient map. Then the following conditions are equivalent:*

(1) *G acts σ-properly on Ω.*
(2) *If $(q(x_n))_{n \in \mathbb{N}}$ is any sequence in Ω/G converging to some $q(x) \in \Omega/G$, then there exists a subsequence $(q(x_{n_m}))_{m \in \mathbb{N}}$ of $(q(x_n))_{n \in \mathbb{N}}$ such that $q^{-1}(M)$ is a σ-trivial G-space, where $M = \overline{\{q(x_{n_m}); m \in \mathbb{N}\}} \subseteq \Omega/G$.*

PROOF. Suppose that the action of G on Ω is σ-proper and let $(q(x_n))_{n \in \mathbb{N}}$ be a sequence in Ω/G converging to $q(x)$. If there exists a constant subsequence $(q(x_{n_m}))_{m \in \mathbb{N}}$ of $(q(x_n))_{n \in \mathbb{N}}$, then (2) follows trivially by taking this subsequence. Otherwise we can find a subsequence $(q(x_{n_m}))_{m \in \mathbb{N}}$ of $(q(x_n))_{n \in \mathbb{N}}$ such that $q(x_{n_m}) \neq q(x)$ for all $m \in \mathbb{N}$ and $q(x_{n_m}) \neq q(x_{n_l})$ if $m \neq l$. By passing to another subsequence if necessary, we may also assume, by the definition of the quotient topology, that $x_{n_m} \to x$ in Ω.

Let M denote the closure of $\{q(x_{n_m}); m \in \mathbb{N}\}$. Then $M = \{q(x_{n_m}); m \in \mathbb{N}\} \cup \{q(x)\}$, since Ω/G is Hausdorff by the σ-properness of the action. It follows that $\mathfrak{s} : M \to q^{-1}(M)$ defined by $\mathfrak{s}(q(x_{n_m})) = x_{n_m}$ and $\mathfrak{s}(q(x)) = x$ is a continuous section. This implies (2).

In order to prove (2) $\Rightarrow$ (1) let $C \subseteq \Omega \times \Omega$ be any compact set and let $(x_n, \dot{s}_n)$ be a sequence in $p^{-1}(C) \subseteq \Omega \times_S G$. Since $(x_n, s_n x_n) \in C$ for all $n \in \mathbb{N}$ we may assume, by passing to a subsequence if necessary, that $(x_n, s_n x_n) \to (x, y)$ for some $(x, y) \in C$. By passing to another subsequence if necessary, we may assume by (2) that $q^{-1}(M)$ is σ-trivial, for $M = \overline{\{q(x_n); n \in \mathbb{N}\}}$. In particular, this implies that $q^{-1}(M)$ is σ-proper. Since $\{(x_n, s_n x_n); n \in \mathbb{N}\} \cup \{(x, y)\} \subseteq q^{-1}(M) \times q^{-1}(M)$ is compact, it follows now that $p^{-1}(\{(x_n, s_n x_n); n \in \mathbb{N}\} \cup \{(x, y)\}) \subseteq q^{-1}(M) \times_S G \subseteq \Omega \times_S G$ is compact, too. Thus $((x_n, \dot{s}_n))_{n \in \mathbb{N}}$ has a convergent subsequence, which finally implies that $p^{-1}(C)$ is compact. $\square$

A first hint that Proposition 5.1.3 is helpful in the investigation of crossed products with continuous trace is given by the following characterization of separable C^*-algebras with continuous trace as those C^*-algebras which have continuous trace on "closures of subsequences". Before we state this characterization let us recall that the *Pedersen ideal* $\mathfrak{m}_A$ of a C^*-algebra A is the minimal dense $*$-ideal of A. It follows directly from the definitions that A has continuous trace if and only if every element of $\mathfrak{m}_A^+$ is a continuous trace element of A.

PROPOSITION 5.1.4. *Let A be a separable C^*-algebra. Then the following statements are equivalent:*

(1) *A has continuous trace.*

(2) *For any convergent sequence $(\rho_n)_{n \in \mathbb{N}}$ in $\widehat{A}$ there exists a subsequence $(\rho_{n_m})_{m \in \mathbb{N}}$ of $(\rho_n)_{n \in \mathbb{N}}$ such that A_M has continuous trace, where M denotes the closure of $\{\rho_{n_m}; m \in \mathbb{N}\}$ in $\widehat{A}$.*

PROOF. It is clear that (1) $\Rightarrow$ (2). Assume that (2) holds. Since by [49] the image of the Pedersen ideal $\mathfrak{m}_A$ in any quotient A/I of A is equal to $\mathfrak{m}_{A/I}$, it follows from (2) applied on constant sequences that $\rho(a)$ is finite for any $a \in \mathfrak{m}_A^+$. So let (ρ_n) be a sequence in $\widehat{A}$ converging to some $\rho \in \widehat{A}$. We have to show that $\mathrm{Tr}\, \rho_n(a) \to \mathrm{Tr}\, \rho(a)$ for all $a \in \mathfrak{m}_A^+$. Applying (2) to any subsequence of $(\rho_n)_{n \in \mathbb{N}}$ we obtain the existence of a subsequence of this subsequence, say $(\rho_m)_{m \in \mathbb{N}}$, such that A_M has continuous trace, if M denotes the closure of $\{\rho_m; m \in \mathbb{N}\}$. Again using the fact that the quotient map $A \to A_M$ maps $\mathfrak{m}_A$ onto $\mathfrak{m}_{A_M}$, this implies that $\mathrm{Tr}\, \rho_m(a) \to \mathrm{Tr}\, \rho(a)$, which finishes the proof. $\square$

We are now going to recall the notion of σ-proper regularizations as introduced in [13].

DEFINITION 5.1.5. Suppose that (A, G, α, τ) is a twisted covariant system and let Ω be a σ-proper G-space. Suppose further that $R : \widehat{A} \to \Omega$ is a continuous, open and surjective G-equivariant map. Then R is called an *open σ-proper regularization* of (A, G, α, τ). If, moreover, Ω is a (locally) σ-trivial G-space, then R is called an *open (locally) σ-trivial regularization* of (A, G, α, τ).

Note that if R is an open σ-proper regularization of (A, G, α, τ) and $S : \Omega \to \mathfrak{K}(G)$ denotes the stabilizer map, then (R, S) is an open pair of regular maps for (A, G, α, τ) with base Ω, so that we can form the subgroup crossed product $A \rtimes_{\alpha,\tau} \Omega^S$ (see Section 4.2). Here we are mainly interested in the case where A has continuous trace, $\Omega = \widehat{A}$, and R is the identity. The following theorem is a very general version of the result of Raeburn and Rosenberg which says that $A \rtimes_\alpha G$ has continuous trace whenever A has continuous trace and $\widehat{A}$ is a free and proper G-space (compare with [**13**, Theorem 4]).

THEOREM 5.1.6. *Suppose that $R : \widehat{A} \to \Omega$ is an open σ-proper regularization for the separable twisted covariant system (A, G, α, τ). Then $A \rtimes_{\alpha,\tau} G$ has continuous trace if and only if $A \rtimes_{\alpha,\tau} \Omega^S$ has continuous trace.*

The proof depends heavily on the following result which is [**13**, Theorem 2].

PROPOSITION 5.1.7. *Suppose that $R : \widehat{A} \to \Omega$ is an open σ-trivial regularization for (A, G, α, τ), and let $\Lambda \subseteq \Omega$ be the image of some continuous section $\mathfrak{s} : \Omega/G \to \Omega$. Then $A \rtimes_{\alpha,\tau} G$ is Morita equivalent to $A_{R^{-1}(\Lambda)} \rtimes_{\alpha,\tau} \Lambda^S$, where $S : \Lambda \to \mathfrak{K}(G)$ denotes the restriction of the stabilizer map to Λ.*

Note that the formulation of the theorem above differs slightly from [**13**, Theorem 2] which says that $A \rtimes_{\alpha,\tau} G$ is Morita equivalent to $C^*(\Lambda^S, A, \alpha, \tau)/I_\Lambda$, where I_Λ denotes the intersection of all kernels of elements of $A_\Lambda \rtimes_{\alpha,\tau} \Lambda^S$. But it is an easy consequence of Proposition 4.2.3 that $A_\Lambda \rtimes_{\alpha,\tau} \Lambda^S$ is isomorphic to $C^*(\Lambda^S, A, \alpha, \tau)/I_\Lambda$.

For the proof of Theorem 5.1.6 it is also convenient to use the following consequence of Theorem 4.1.6 and [**13**, Theorem 1].

PROPOSITION 5.1.8. *Let $R : \widehat{A} \to \Omega$ be an open σ-proper regularization of (A, G, α, τ). Then the map*

$$\text{ind} : (A \rtimes_{\alpha,\tau} \Omega^S)\widehat{\ }/G \to (A \rtimes_{\alpha,\tau} G)\widehat{\ }; \ G((x, \rho \times V)) \mapsto \text{ind}_{S_x}^G(\rho \times V)$$

is a homeomorphism onto $(A \rtimes_{\alpha,\tau} G)\widehat{\ }$.

PROOF OF THEOREM 5.1.6. Suppose first that $A \rtimes_{\alpha,\tau} \Omega^S$ has continuous trace and let $(\pi_n \times U^n)_{n \in \mathbb{N}}$ be a sequence in $(A \rtimes_{\alpha,\tau} G)\widehat{\ }$ which converges to some $\pi \times U \in (A \rtimes_{\alpha,\tau} G)\widehat{\ }$. Then, by Proposition 5.1.8 and by passing to a subsequence if necessary, we may find a sequence $(x_n, \rho_n \times V^n)_{n \in \mathbb{N}} \subseteq (A \rtimes_{\alpha,\tau} \Omega^S)\widehat{\ }$ converging to some $(x, \rho \times V) \in (A \rtimes_{\alpha,\tau} \Omega^S)\widehat{\ }$ such that $\pi_n \times U^n = \text{ind}_{S_x}^G(\rho_n \times V^n)$ and $\pi \times U = \text{ind}_{S_x}^G(\rho \times V)$. This implies that $x_n \to x$ in Ω and that $q(x_n) \to q(x)$ in Ω/G, where $q : \Omega \to \Omega/G$ denotes the quotient map.

Since Ω is a σ-proper G-space it follows from Proposition 5.1.3 that there exists a subsequence $(x_{n_m})_{m \in \mathbb{N}}$ of $(x_n)_{n \in \mathbb{N}}$ such that $q^{-1}(M)$ is σ-trivial for

$M = \{q(x_{n_m}); m \in \mathbb{N}\} \cup \{q(x)\}$. Let $\Lambda \subseteq q^{-1}(M)$ be the image of a continuous section. Then it follows from Proposition 5.1.7 that $A_{R^{-1}(q^{-1}(M))} \rtimes_{\alpha,\tau} G$ is Morita equivalent to $A_{R^{-1}(\Lambda)} \rtimes_{\alpha,\tau} \Lambda^S$. But $A_{R^{-1}(\Lambda)} \rtimes_{\alpha,\tau} \Lambda^S$ has continuous trace since it is a quotient of the continuous trace algebra $A \rtimes_{\alpha,\tau} \Omega^S$ (see Proposition 4.2.4). Thus $A_{R^{-1}(q^{-1}(M))} \rtimes_{\alpha,\tau} G$ has continuous trace, too [70, Theorem 2.14]. Now, if $(\pi_{n_m} \times U^{n_m})_{m \in \mathbb{N}}$ is the corresponding subsequence of $(\pi_n \times U^n)_{n \in \mathbb{N}}$, then the closure of this subsequence is contained in $(A_{R^{-1}(q^{-1}(M))} \rtimes_{\alpha,\tau} G)\hat{\ }$, since for all $m \in \mathbb{N}$

$$\ker(\pi_{n_m} \times U^{n_m}) = \ker(\mathrm{ind}_{S_{x_{n_m}}}^G (\rho_{n_m} \times V^{n_m})) = \bigcap_{s \in G} \ker \rho_{n_m} \circ \alpha_s$$

$$\supseteq \ker R^{-1}(G(x_{n_m})) \supseteq \ker R^{-1}(q^{-1}(M)).$$

But this implies that $A \rtimes_{\alpha,\tau} G$ has continuous trace by Proposition 5.1.4.

Suppose now that $A \rtimes_{\alpha,\tau} G$ has continuous trace, and assume that $(x_n, \rho_n \times V^n)_{n \in \mathbb{N}}$ is a sequence in $(A \rtimes_{\alpha,\tau} \Omega^S)\hat{\ }$ which converges to some $(x, \rho \times V) \in (A \rtimes_{\alpha,\tau} \Omega^S)\hat{\ }$. As above we find a subsequence $(q(x_{n_m}))_{m \in \mathbb{N}}$ of $(q(x_n))_{n \in \mathbb{N}}$ such that $q^{-1}(M)$ is σ-trivial, if $M = \{q(x_{n_m}); m \in \mathbb{N}\} \cup \{q(x)\}$. By passing to another subsequence if necessary, we may assume one of the following situations

(1) $q(x_{n_m}) \neq q(x_{n_l})$ if $m \neq l$ and $q(x_{n_m}) \neq q(x)$ for all $m \in \mathbb{N}$, or

(2) $q(x_{n_m}) = q(x)$ for all $m \in \mathbb{N}$.

In the first situation we conclude that $\Lambda = \{x_{n_m}; m \in \mathbb{N}\} \cup \{x\}$ is the image of a continuous section for $q^{-1}(M)$, from which follows that $A_{R^{-1}(\Lambda)} \rtimes_{\alpha,\tau} \Lambda^S$ is Morita equivalent to $A_{R^{-1}(q^{-1}(M))} \rtimes_{\alpha,\tau} G$. But we know by assumption that $A_{R^{-1}(q^{-1}(M))} \rtimes_{\alpha,\tau} G$ has continuous trace, which implies that $A_{R^{-1}(\Lambda)} \rtimes_{\alpha,\tau} \Lambda^S$ has continuous trace, too. Thus, the subsequence $(x_{n_m}, \rho_{n_m} \times V^{n_m})_{m \in \mathbb{N}}$ has its closure in the dual of the continuous trace algebra $A_{R^{-1}(\Lambda)} \rtimes_{\alpha,\tau} \Lambda^S$.

In Situation (2) we see that the closure of the sequence $(x_{n_m}, \rho_{n_m} \times V^{n_m})_{m \in \mathbb{N}}$ lies in the dual of $A_{R^{-1}(G(x))} \rtimes_{\alpha,\tau} G(x)^S$. Thus we are done if we can show that this algebra has continuous trace. For this observe that the map

$$\phi : (A_{R^{-1}(G(x))} \rtimes_{\alpha,\tau} G(x)^S)\hat{\ } \to G/S_x; (sx, \sigma \times W) \to sS_x$$

is a continuous G-equivariant map such that $A_{R^{-1}(G(x))} \rtimes_{\alpha,\tau} G(x)^S/I \cong A_x \rtimes_{\alpha^x,\tau^x} S^x$, if $I = \ker \phi^{-1}(eS_x)$. Thus it follows from [10] (see also [13, Theorem 3]) that $A_{R^{-1}(G(x))} \rtimes_{\alpha,\tau} G(x)^S$ is isomorphic to the induced C^*-algebra $\mathrm{Ind}_{S_x}^G (A_x \rtimes_{\alpha^x,\tau^x} S_x)$. Since $\mathrm{Ind}_{S_x}^G (A_x \rtimes_{\alpha^x,\tau^x} S_x)$ has continuous trace if and only if $A_x \rtimes_{\alpha^x,\tau^x} S_x$ has continuous trace [13, Proposition 10], it remains to show that $A_x \rtimes_{\alpha^x,\tau^x} S_x$ has continuous trace. But this follows from the fact that $A_x \rtimes_{\alpha^x,\tau^x} S_x$ is Morita equivalent to $A_{R^{-1}(G(x))} \rtimes_{\alpha,\tau} G$ which in turn is a quotient of the continuous trace algebra $A \rtimes_{\alpha,\tau} G$. $\square$

As a direct corollary we obtain a part of Williams's description of transformation groups which have continuous trace transformation group algebras [70, Theorem 2.7].

COROLLARY 5.1.9. *Suppose that (G, Ω) is a second countable locally compact transformation group such that G acts σ-properly on Ω. Then the transformation group algebra $C^*(G,\Omega) = C_0(\Omega) \rtimes G$ has continuous trace if and only if $C^*(\Omega^S)$ has continuous trace, where $S : \Omega \to \mathfrak{K}(G)$ denotes the stabilizer map.*

5.2. Pointwise unitary subgroup actions

As a second step towards our general description of separable abelian systems having continuous-trace twisted crossed products we want to investigate the structure of subgroup crossed products $A \rtimes_{\alpha,\tau} \Omega^S$ in the case where Ω^S acts pointwise unitarily on the continuous trace C^*-algebra A.

DEFINITION 5.2.1. Suppose that A is a C^*-algebra with continuous trace. Let $\Omega = \hat{A}$ and let $p : E \to \Omega$ be a C^*-bundle with $A = \Gamma_0(E)$. Assume further that $H : \Omega \to \mathfrak{K}(G)$ is a continuous map and that (α, τ) is a twisted subgroup action of Ω^H on A. Then (α, τ) is called *unitary* if there exists a map $u : \Omega^H \to \prod_{x \in \Omega} \mathcal{U}(A_x)$ which satisfies the following properties:

    (1) $u(x,s) \in \mathcal{U}(A_x)$ and the maps $\Omega^H \to E$, $(x,s) \mapsto u(x,s)a(x)$ and $(x,s) \mapsto a(x)u(x,s)$ are continuous for all $a \in A$,

    (2) $u(x, \cdot)$ is a homomorphism from H_x into $\mathcal{U}(A_x)$ for each $x \in \Omega$,

    (3) $\alpha_s^x(a(x)) = u(x,s)a(x)u(x,s)^*$ for all $s \in H_x$, and

    (4) $u(x,n) = \tau_n^x$ for all $x \in \Omega$ and $n \in N_\tau$.

If u is as above, then u is called a *unitary which implements* (α, τ). Moreover, (α, τ) is called *locally unitary*, if each $x \in \Omega$ has an open neighborhood V such that (α^V, τ^V) is unitary. Finally, (α, τ) is called *pointwise unitary* if for each $x \in \Omega$ there exists a strictly continuous homomorphism $U^x : H_x \to \mathcal{U}(A_x)$ such that (ρ_x, U^x) becomes a covariant representation of $(A_x, H_x, \alpha^x, \tau^x)$, where ρ_x denotes evaluation at x.

Note that it follows from conditions (1), (3) and (4) of the definition above that $(\rho_x, \rho_x \circ u(x, \cdot))$ is a covariant representation of $(A_x, H_x, \alpha^x, \tau^x)$ into $\mathcal{M}(A_x)$ for each $x \in \Omega$. This shows that locally unitary twisted subgroup actions are always pointwise unitary. Locally unitary subgroup actions were first defined by Raeburn and Williams in [56], and have been investigated also in [13]. We will now see that subgroup crossed products by unitary twisted subgroup actions have a very easy description (compare with the proofs of [13, Lemma 2 and Theorem 5]).

PROPOSITION 5.2.2. *Suppose that (α, τ) is a unitary twisted subgroup action of Ω^H on the continuous trace C^*-algebra A. Let $\tilde{G} = G/N_\tau$ and let $\tilde{H} : \Omega \to \mathfrak{K}(\tilde{G})$ denote the continuous map given by $\tilde{H}_x = H_x/N_\tau$. Then $A \rtimes_{\alpha,\tau} \Omega^H$ is isomorphic to the quotient of $A \otimes C^*(\Omega^{\tilde{H}})$ by the ideal*

$$I = \bigcap \{\ker(\rho_x \otimes (x, U)); x \in \Omega, U \in \widehat{\tilde{H}_x}\},$$

where ρ_x denotes any representative of $x \in \Omega = \hat{A}$. In particular, $(A \rtimes_{\alpha,\tau} \Omega^H)^\frown$ is homeomorphic to $C^*(\Omega^{\tilde{H}})^\frown$ and $A \rtimes_{\alpha,\tau} \Omega^H$ has continuous trace if and only if

$C^*(\Omega^{\tilde{H}})$ has continuous trace.

PROOF. Define $\Phi : A \odot C_c(\Omega^{\tilde{H}}) \to \Gamma_c(q^*E, \tau)$ by

$$\Phi(a \otimes f)(x, s) = a(x)f(x, \tilde{s})u(x, s)^*,$$

where $u : \Omega^H \to \prod_{x \in \Omega} \mathcal{U}(A_x)$ is a unitary which implements (α, τ). It follows easily from [56, Lemma 2.1] that the image of $A \odot C_c(\Omega^{\tilde{H}})$ is dense in $A \rtimes_{\alpha, \tau} \Omega^H$, and easy computations show that

$$\rho_x \otimes (x, U)(a \otimes f) = (x, U \otimes (\rho_x \times \rho_x \circ u(x, \cdot)))(\Phi(a \otimes f))$$

for all $x \in \Omega$, $U \in \widehat{\tilde{H}}$ and all elementary tensors $f \otimes a \in A \odot C_c(\Omega^{\tilde{H}})$. Since for each $x \in \Omega$ the set $\{U \otimes (\rho_x \times \rho_x \circ u(x, \cdot)); U \in \widehat{\tilde{H}}\}$ equals $(A_x \rtimes_{\alpha^x, \tau^x} H_x)\widehat{}$ by the second step of the Mackey machine, we conclude that Φ extends to a *-homomorphism from $A \otimes C^*(\Omega^{\tilde{H}})$ onto $A \rtimes_{\alpha, \tau} \Omega^H$ with kernel I. In particular, we see that $(A \rtimes_{\alpha, \tau} \Omega^H)\widehat{}$ is homeomorphic to $C^*(\Omega^{\tilde{H}})\widehat{}$.

Since the property of having continuous trace is inherited by tensor products and quotients, it follows that $A \rtimes_{\alpha, \tau} \Omega^H$ has continuous trace if the same is true for $C^*(\Omega^{\tilde{H}})$. For the opposite direction assume that $A \rtimes_{\alpha, \tau} \Omega^H \cong (A \otimes C^*(\Omega^{\tilde{H}}))/I$ has continuous trace. To see that $C^*(\Omega^{\tilde{H}})$ has continuous trace as well, let f and a be positive elements in the Pedersen ideals $\mathfrak{m}_{C^*(\Omega^{\tilde{H}})}$ and $\mathfrak{m}_A$ of $C^*(\Omega^{\tilde{H}})$ and A, respectively. Then $f \otimes a$ is in the Pedersen ideal of $A \otimes C^*(\Omega^{\tilde{H}})$. This follows almost directly from the definition of the Pedersen ideal as given in [50, page 175]. Since the Pedersen ideal is carried over by quotient maps (see [49]), it follows that $(f \otimes a) + I$ lies in $\mathfrak{m}_{(A \otimes C^*(\Omega^{\tilde{H}})/I}$. Assume now that $(x, U) \mapsto \mathrm{tr}((x, U)(f))$ fails to be finite or continuous at a point $(x_0, U^0) \in C^*(\Omega^{\tilde{H}})\widehat{}$. Choose $a \in \mathfrak{m}_A$ such that $\mathrm{tr}(\rho_{x_0}(a)) > 0$. Then the map $((A \otimes C^*(\Omega^{\tilde{H}})/I)\widehat{} \to \mathbb{C}$ given by

$$\rho_x \otimes (x, U) \mapsto \mathrm{tr}(\rho_x \otimes (x, U)(a \otimes f)) = \mathrm{tr}(\rho_x(a))\,\mathrm{tr}((x, U)(f))$$

is finite and continuous by assumption. Since A has continuous trace, $x \mapsto \mathrm{tr}(\rho_x(a))$ is finite and continuous, too. Since $\mathrm{tr}(\rho_{x_0}(a)) \neq 0$, this clearly contradicts our assumption on (x_0, U^0). Thus we conclude that $C^*(\Omega^{\tilde{H}})$ has continuous trace. $\square$

As an easy consequence of the above result and Theorem 5.1.6 we obtain the following result which extends [13, Theorem 5]

THEOREM 5.2.3. *Suppose that (A, G, α, τ) is a twisted covariant system such that A has continuous trace and such that the action of G on $\Omega = \widehat{A}$ is σ-proper. Suppose further that (α, τ) is locally unitary on the stabilizers, which means that the restriction of (α, τ) to Ω^S is locally unitary, where $S : \Omega \to \mathfrak{K}(G)$ denotes the stabilizer map. Then the following conditions are equivalent.*

(1) *$A \rtimes_{\alpha, \tau} G$ has continuous trace.*
(2) *$C^*(\Omega^{\tilde{S}})$ has continuous trace*
(3) *The transformation group algebra $C^*(\tilde{G}, \Omega)$ has continuous trace.*

As usual $\tilde{S}_x = S_x/N_\tau$ and $\tilde{G} = G/N_\tau$.

PROOF. By Theorem 5.1.6 we know that $A \rtimes_{\alpha,\tau} G$ has continuous trace if and only if $A \rtimes_{\alpha,\tau} \Omega^S$ has continuous trace. By assumption (α, τ) is a locally unitary action of Ω^S on A. Thus for any $x \in \Omega$ we find a neighborhood W of x such that the restriction of (α, τ) to W^S is unitary. Hence by Proposition 5.2.2 we know that $A_W \rtimes_{\alpha,\tau} W^S$ has continuous trace if and only if $C^*(W^{\tilde{S}})$ has continuous trace. Since having continuous trace is a local property, this easily implies that $A \rtimes_{\alpha,\tau} \Omega^S$ has continuous trace if and only if $C^*(\Omega^{\tilde{S}})$ has continuous trace. Thus we conclude $(1) \Leftrightarrow (2)$. The equivalence of (2) and (3) is Corollary 5.1.9 (or Williams's theorem [**70**, Theorem 2.7]). $\square$

If G is abelian and $H : \Omega \to \mathfrak{K}(G)$ is a continuous map, then we know from Lemma 4.4.4 that $C^*(\Omega^H)$ is isomorphic to $C_0(\Omega \times_{H^\perp} \hat{G})$. It is easily seen that this isomorphism is $\hat{G}$-equivariant with respect to the canonical $\hat{G}$-actions on $C^*(\Omega^H)$ and $C_0(\Omega \times_{H^\perp} \hat{G})$. Thus, as a direct consequence of Proposition 5.2.2 we obtain the following result, which is also implicitly contained in [**56**, Proposition 6.1].

PROPOSITION 5.2.4. *Suppose that A is a continuous trace algebra with dual space $\hat{A} = \Omega$, $H : \Omega \to \mathfrak{K}(G)$ is a continuous map, and (α, τ) is a unitary twisted subgroup action of Ω^H on A. If $\tilde{G} = G/N_\tau$ is abelian, then $A \rtimes_{\alpha,\tau} \Omega^H$ has continuous trace and $(A \rtimes_{\alpha,\tau} \Omega^H)\hat{}$ is homeomorphic to $\Omega \times_{H^\perp} \hat{\tilde{G}}$ as a $\hat{\tilde{G}}$-space.*

In case where α is a pointwise unitary action of a separable locally compact group G on the continuous trace algebra A, it has been shown by Rosenberg in [**64**] that α is automatically locally unitary if G satisfies some additional conditions on its structure. One sufficient condition is, for instance, that G is compactly generated abelian. This result has a direct connection to Palais's slice theorem, via the duality between locally unitary actions of an abelian group G on A and the local triviality of the proper $\hat{G}$-space $(A \rtimes_\alpha G)\hat{}$ (see the proof of Theorem 6.1.4 below). As for the slice theorem, there is no analogue of Rosenberg's theorem for pointwise unitary subgroup actions, which is the main reason that Olesen's and Raeburn's proof of [**46**, Theorem 1.10] does not carry over to the case of pointwise unitary subgroup actions. But in analogy with the fact that all σ-proper actions are "σ-trivial on closures of subsequences" (see Proposition 5.1.3) we will now see that, at least if G is abelian, pointwise unitary actions are "unitary on closures of subsequences".

PROPOSITION 5.2.5. *Suppose that A is a separable stable continuous trace algebra with dual space $\hat{A} = \Omega$ and let $H : \Omega \to \mathfrak{K}(G); x \mapsto H_x$ be a continuous map, for some separable locally compact group G, such that H_x is abelian for all $x \in \Omega$. Assume further that α is a pointwise unitary subgroup action of Ω^H on A. Then the following is true: If $(x_n)_{n\in\mathbb{N}}$ is a sequence in Ω such that $x_n \to x$ for some $x \in \Omega$, then there exists a subsequence $(x_{n_m})_{m\in\mathbb{N}}$ such that for*

$M = \{x_{n_m}; m \in \mathbb{N}\} \cup \{x\}$ the restriction of α to M^H is unitary. Moreover, the subgroup crossed product $A \rtimes_\alpha \Omega^H$ has continuous trace.

Unfortunately, the proof of this proposition is much more complicated than the proof of its counterpart for σ-proper actions. We start with an easy lemma. We omit the straightforward proof.

LEMMA 5.2.6. Suppose that α is a subgroup action of Ω^H on $A = \Gamma_0(E)$. Then there are canonical $*$-homomorphisms $i_{\Omega^H} : C^*(\Omega^H) \to \mathcal{M}(A \rtimes_\alpha \Omega^H)$ and $i_A : A \to \mathcal{M}(A \rtimes_\alpha \Omega^H)$ which are defined by the equations

$$i_{\Omega^H}(f)g(x,s) = \int_{H_x} f(x,t)\alpha_t(g(x,t^{-1}s))\,d_{H_x}t$$

and

$$i_A(a)g(x,s) = a(x)g(x,s),$$

for $f \in C_c(\Omega^H)$, $a \in A$ and $g \in \Gamma_c(q^*E)$. Furthermore, if $(x, \pi \times U)$ is a subgroup representation of $A \rtimes_\alpha \Omega^H$, then $(x, \pi \times U)(i_{\Omega^H}(f)) = (x,U)(f)$ for all $f \in C^*(\Omega^H)$ and $(x, \pi \times U)(i_A(a)) = \pi(a(x))$ for all $a \in A$.

For any C^*-algebra A and Hilbert space $\mathcal{H}$ we denote by $\text{Rep}(A, \mathcal{H})$ the set of all $*$-representations of A into $\mathcal{H}$ equipped with the strong topology, i.e. a net $(\pi_i)_{i \in I}$ converges to π in this topology if and only if $\pi_i(a)\xi \to \pi(a)\xi$ for all $a \in A$ and $\xi \in \mathcal{H}$. If A and $\mathcal{H}$ are separable, then $\text{Rep}(A, \mathcal{H})$ is a polish space [9, Proposition 3.7.1], which allows to restrict ourselves to sequences. The natural map from $\text{Rep}(A, \mathcal{H})$ to $\text{Rep}(A)$ which maps a representation π onto its equivalence class is always continuous [22, Lemma 2.3]. Moreover, if we denote the set of irreducible elements of $\text{Rep}(A, \mathcal{H})$ by $\text{Irr}(A, \mathcal{H})$, then the map from $\text{Irr}(A, \mathcal{H})$ into $\widehat{A}$ which maps an element of $\text{Irr}(A, \mathcal{H})$ to its equivalence class is continuous and open by [9, Theorem 3.5.8]. It follows from trivial calculations that the natural map from $\text{Irr}(A, \mathcal{H}) \to \text{Irr}(\mathcal{M}(A), \mathcal{H})$ which is given by extension of representations is a homeomorphism onto its image.

Suppose now that $H : \Omega \to \mathfrak{K}(G)$ is a continuous map. We denote by $\mathcal{S}(\Omega^H, \mathcal{H})$ the set of all subgroup representations of $C^*(\Omega^H)$ on $\mathcal{H}$, equipped with the relative topology of $\text{Rep}(C^*(\Omega^H), \mathcal{H})$.

LEMMA 5.2.7. Suppose that G, $\mathcal{H}$ and Ω are separable and let $(x_n)_{n \in \mathbb{N}}$ be a sequence in Ω converging to some $x \in \Omega$ such that $x_n \neq x_m$ for all $n \neq m$ and $x_n \neq x$ for all $n \in \mathbb{N}$. Suppose further that the sequence $(x_n, U^{x_n})_{n \in \mathbb{N}} \subseteq \mathcal{S}(\Omega^H, \mathcal{H})$ converges to some element $(x, U^x) \in \mathcal{S}(\Omega^H, \mathcal{H})$. Let $M = \{x_n; n \in \mathbb{N}\} \cup \{x\}$. Then the map $u : M^H \to \mathcal{U}(\mathcal{H})$ defined by $u(y,s) = U_s^y$ is strongly continuous.

PROOF. It is clearly enough to show that $U_{s_n}^{x_n} \to U_s^x$ in the strong topology whenever $s_n \in H_{x_n}$ and $s \in H_x$ such that $s_n \to s$ in G. For this let $f \in C_c(\Omega^H)$ and choose $F \in C_c(\Omega \times G)$ such that $F|_{\Omega^H} = f$. For $s \in G$ define $F_s \in C_c(\Omega \times G)$ by $F_s(x,t) = F(x, s^{-1}t)$. Then the map $G \to C_c(\Omega \times G); s \mapsto F_s$

is continuous with respect to the inductive limit topology on $C_c(\Omega \times G)$, which implies that the map $G \to C_c(\Omega^H); s \mapsto f_s = F_s|_{\Omega^H}$ is continuous with respect to the inductive limit topology on $C_c(\Omega^H)$, and hence also with respect to the C^*-norm on $C_c(\Omega^H)$.

Thus, if $(s_n)_{n \in \mathbb{N}}$ and s are as above, then we conclude easily that

$$U_{s_n}^{x_n}(x_n, U^{x_n})(f)\xi = (x_n, U^{x_n})(f_{s_n})\xi \to (x, U^x)(f_s)\xi = U_s^x(x, U^x)(f)\xi$$

for all $\xi \in \mathcal{H}$. Taking $\eta = (x, U^x)(f)\xi$ we now deduce that

$$
\begin{aligned}
\|U_{s_n}^{x_n}\eta - U_s^x\eta\| &= \|U_{s_n}^{x_n}(x, U^x)(f)\xi - U_s^x(x, U^x)(f)\xi\| \\
&\leq \|U_{s_n}^{x_n}(x, U^x)(f)\xi - U_{s_n}^{x_n}(x_n, U^{x_n})(f)\xi\| \\
&\quad + \|U_{s_n}^{x_n}(x_n, U^{x_n})(f)\xi - U_s^x(x, U^x)(f)\xi\|
\end{aligned}
$$

converges to zero. The proof follows now from the fact that $\{(x, U^x)(f)\xi; f \in C_c(\Omega^H), \xi \in \mathcal{H}\}$ is dense in $\mathcal{H}$. $\square$

The next lemma is certainly well known, but for completeness we include the short proof.

LEMMA 5.2.8. *Let $\mathcal{K}$ denote the compact operators on the separable Hilbert space $\mathcal{H}$, and suppose that $(u_n)_{n \in \mathbb{N}}$ is a sequence of unitary operators on $\mathcal{H}$ such that $u_n T u_n^*$ converges strongly to T for all $T \in \mathcal{K}$. Then $u_n T u_n^*$ converges to T in the operator norm for all $T \in \mathcal{K}$.*

PROOF. Let $\xi \in \mathcal{H}$ with $\|\xi\| = 1$ and let P_ξ denote the projection onto $\mathbb{C}\xi$. Then $u_n P_\xi u_n^*$ is the projection onto $u_n \xi$, and for each $\eta \in \mathcal{H}$ we obtain

$$\langle \eta, u_n\xi \rangle u_n\xi = P_{u_n\xi}\eta \to P_\xi\eta = \langle \eta, \xi \rangle \xi.$$

Choosing $\eta = \xi$, it follows in particular that

$$|\langle \xi, u_n\xi \rangle| \to |\langle \xi, \xi \rangle| = 1.$$

Thus we may find a sequence $(t_n)_{n \in \mathbb{N}} \subseteq \mathbb{T}$ such that $\langle \xi, u_n t_n \xi \rangle \to 1$, from which then follows that $u_n t_n \xi \to \xi$ in $\mathcal{H}$. But it follows easily from this that $P_{u_n\xi} = P_{u_n t_n \xi} \to P_{u\xi}$ in $\mathcal{K}$. Since $\mathcal{K}$ is generated as an algebra by the projections of rank one, it follows now that $u_n T u_n^* \to T$ in the operator norm for every $T \in \mathcal{K}$. This finishes the proof. $\square$

We are now ready to give the

PROOF OF PROPOSITION 5.2.5. If $(x_n)_{n \in \mathbb{N}}$ contains a constant subsequence, then we pass over to this subsequence and we are done. Otherwise, also by passing to a subsequence if necessary, we may assume that $x_n \neq x_m$ for all $n \neq m$ and $x_n \neq x$ for all $n \in \mathbb{N}$. Since A is a separable stable continuous trace algebra there is a neighborhood V of x in Ω such that A_V is isomorphic to $C_0(V, \mathcal{K})$, where $\mathcal{K}$ denotes the algebra of compact operators on the separable infinite dimensional Hilbert space $\mathcal{H}$. Restricting α to V^H, we may as well assume that $V = \Omega$ and $A = C_0(\Omega, \mathcal{K})$.

For each $y \in \Omega$ let $\rho_y : A \to \mathcal{K}$ denote evaluation at y. Since the Mackey obstructions of the systems $\mathcal{K} \rtimes_{\alpha^y} H_y$ are trivial by assumption for all $y \in \Omega$, we conclude that for each $y \in \Omega$ there exists a unitary representation U^y of H_y on $\mathcal{H}$ such that (ρ_y, U^y) is a covariant representation of $\mathcal{K} \rtimes_{\alpha^y} H_y$. This implies that a complete set of representatives for $(A \rtimes_\alpha \Omega^H)\widehat{\;}$ is given by the set of representations

$$\mathcal{S} = \{\rho_y \times \chi U^y; y \in \Omega, \chi \in \widehat{H}_y\}.$$

Note that all representations in $\mathcal{S}$ act on the fixed Hilbert space $\mathcal{H}$.

Since by Proposition 4.2.5 the projection from $(A \rtimes_\alpha \Omega^H)\widehat{\;}$ onto Ω is open, we may assume, by passing to another subsequence if necessary, that $(x_n, \rho_{x_n} \times U^{x_n})$ converges to $(x, \rho_x \times U^x)$ in $(A \rtimes_\alpha \Omega^H)\widehat{\;}$. Since the natural map from $\mathrm{Irr}(A \rtimes_\alpha \Omega^H, \mathcal{H}) \to (A \rtimes_\alpha \Omega^H)\widehat{\;}$ is also continuous and open we may as well assume, again by passing to a subsequence if necessary, that there is sequence $(u_n)_{n \in \mathbb{N}}$ of unitary operators on $\mathcal{H}$ such that $\mathrm{Ad}\, u_n \circ (x_n, \rho_{x_n} \times U^{x_n}) = (x_n, \mathrm{Ad}\, u_n \circ \rho_{x_n} \times \mathrm{Ad}\, u_n \circ U^{x_n})$ converges to $(x, \rho_x \times U^x)$ in $\mathrm{Irr}(A \rtimes_\alpha \Omega^H, \mathcal{H})$. By the continuity of extending representations to the multiplier algebra, it follows now from Lemma 5.2.6 that $(x_n, \mathrm{Ad}\, u_n \circ U^{x_n})$ converges to (x, U) in $\mathcal{S}(\Omega^H, \mathcal{H})$, and that $\mathrm{Ad}\, u_n \circ \rho_{x_n}$ converges to ρ_x in $\mathrm{Irr}(A, \mathcal{H})$. The second condition especially implies that $u_n T u_n^*$ converges strongly to T for every $T \in \mathcal{K}$. Thus it follows from Lemma 5.2.8 that $\mathrm{Ad}\, u_n \to \mathrm{id}$ in $\mathrm{Aut}(\mathcal{K})$. Since $u \mapsto \mathrm{Ad}\, u$ factors through an isomorphism between $\mathcal{PU}$ and $\mathrm{Aut}(\mathcal{K})$, and since the quotient map $\mathcal{U} \to \mathcal{PU}$ has local sections, we may multiply each u_n with a suitable element of $\mathbb{T}$ in order to assume that u_n converges strongly to $\mathrm{Id}_{\mathcal{H}}$.

Now let $M = \{x_n; n \in \mathbb{N}\} \cup \{x\}$, and for $y \in M$ let $u_y = u_n$ if $y = x_n$ and $u_x = \mathrm{Id}_{\mathcal{H}}$. Then the map $y \mapsto u_y : M \to \mathcal{U}(\mathcal{H})$ is strongly continuous. We define a map $u : M^H \to \mathcal{U}(\mathcal{H})$ by $u(y, s) = U_s^y$, $(y, s) \in M^H$. Since $(x_n, \mathrm{Ad}\, u_n \circ U_n)_{n \in \mathbb{N}}$ converges to (x, U) in $\mathcal{S}(\Omega^H, \mathcal{H})$, it follows from Lemma 5.2.7 that the map $(y, s) \mapsto u_y u(y, s) u_y^*$ is strongly continuous. Thus u is strongly continuous, too. Moreover, it is clear that u implements α in $A_M = C(M, \mathcal{K})$. Thus, since the strong topology on $\mathcal{U}(\mathcal{H}) \cong \mathcal{U}(\mathcal{K})$ coincides with the strict topology, it follows that M^H acts unitarily on A_M.

Let us finally show that $A \rtimes_\alpha \Omega^H$ has continuous trace. For this let $(x_n, \rho_n \times V^n)_{n \in \mathbb{N}}$ be a sequence in $(A \rtimes_\alpha \Omega^H)\widehat{\;}$ which converges to some $(x, \rho \times V) \in (A \rtimes_\alpha \Omega^H)\widehat{\;}$. Then $x_n \to x$ in Ω and, by passing to a subsequence if necessary, we may assume by the first part of the proof that α restricts to a unitary action of M^H to A_M, for $M = \{x_n; n \in \mathbb{N}\} \cup \{x\}$. By Proposition 5.2.2 we know that $A_M \rtimes_\alpha M^H$ has continuous trace if and only if $C^*(M^H)$ has continuous trace, which is trivially true since $C^*(M^H)$ is commutative. Thus the closure of the (sub-) sequence $(x_n, \rho_n \times V^n)_{n \in \mathbb{N}}$ is contained in in the dual of the continuous-trace algebra $A_M \rtimes_\alpha M^H$ which implies that $A \rtimes_\alpha \Omega^H$ has continuous trace by Proposition 5.1.4. $\square$

If A is a C^*-algebra, Λ and Ω are locally compact spaces, and $P : \widehat{A} \to \Omega$ and $f : \Lambda \to \Omega$ are continuous maps, then the pull-back $C_0(\Lambda) \otimes_{C(\Omega)} A$ has been defined as the quotient of $C_0(\Lambda, A)$ by the ideal

$$I_\Omega = \bigcap \{\ker(y, \rho); y \in \Lambda, \rho \in \widehat{A} \text{ such that } f(y) = P(\rho)\}.$$

(See [**55**, §1] for a more general definition of pull-backs of C^*-algebras.) If $A = \Gamma_0(E)$ for a C^*-bundle $p : E \to \Omega$, and if $P : \widehat{A} \to \Omega$ denotes the corresponding map, then $C_0(\Lambda) \otimes_{C(\Omega)} A$ is isomorphic to $\Gamma_0(f^*E)$, where $q : f^*E \to \Lambda$ denotes the pull-back of the C^*-bundle $p : E \to \Omega$ via f as defined in Section 4.2 (see [**55**, Proposition 1.3]).

THEOREM 5.2.9. *Suppose that (A, G, α, τ) is a separable twisted covariant system such that A is a continuous trace algebra with dual space $\widehat{A} = \Omega$. Let $H : \Omega \to \mathfrak{K}(G)$ be a G-equivariant continuous map such that $N_\tau \subseteq H_x \subseteq S_x$, $\tilde{H}_x = H_x/N_\tau$ is abelian for each $x \in \Omega$, and such that the restriction of (α, τ) to Ω^H is pointwise unitary. Then the following are true:*

(1) *$A \rtimes_{\alpha,\tau} \Omega^H$ has continuous trace.*

(2) *$A \rtimes_{\alpha,\tau} \Omega^H$ is isomorphic to the pull-back*

$$\text{res}^* A = C_0((A \rtimes_{\alpha,\tau} \Omega^H)\widehat{\,}) \otimes_{C(\Omega)} A,$$

where $\text{res} : (A \rtimes_{\alpha,\tau} \Omega^H)\widehat{\,} \to \Omega$ denotes the canonical projection.

(3) *If $\tilde{G} = G/N_\tau$ is abelian, then $(A \rtimes_{\alpha,\tau} \Omega^H)\widehat{\,}$ is a σ-proper $\widehat{\tilde{G}}$-space with respect to the dual action $\widehat{\alpha}$ of $\widehat{\tilde{G}}$ on $A \rtimes_{\alpha,\tau} \Omega^H$.*

For the proof of this theorem we need the following lemma, which is an extension of Proposition 5.2.5 for pointwise unitary twisted actions.

LEMMA 5.2.10. *Let (A, G, α, τ) and $H : \Omega \to \mathfrak{K}(G)$ be as in Proposition 5.2.9 above and suppose that $(x_n)_{n\in\mathbb{N}}$ converges to x in Ω. Then there exists a subsequence $(x_{n_m})_{m\in\mathbb{N}}$ such that, for $M = \{x_{n_m}; m \in \mathbb{N}\} \cup \{x\}$, the restriction (α^M, τ^M) of (α, τ) on A_M is unitary. Moreover, $A \rtimes_{\alpha,\tau} \Omega^H$ has continuous trace.*

PROOF. By Theorem 2.1.7 we find an action β of $\tilde{G} = G/N_\tau$ on a C^*-algebra B such that β is Morita equivalent to (α, τ) via an $(\alpha, \tau) - \beta$ imprimitivity pair (X, u). Passing over to the action $\beta \otimes \text{id}$ on $B \otimes \mathcal{K}$, if necessary, we may assume that B is stable. Let us identify $\widehat{B}$ with $\widehat{A} = \Omega$ by the homeomorphism given by induction via X, and let $\tilde{H} : \Omega \to \mathfrak{K}(G)$ be defined by $\tilde{H}_x = H_x/N_\tau$. Then we restrict β to an action of $\Omega^{\tilde{H}}$ on B.

Using Proposition 5.2.5 it is obviously enough to prove the following: If M is any closed subset of Ω such that β^M is a unitary action of $M^{\tilde{H}}$ on B_M, then (α^M, τ^M) is a unitary action of Ω^H on A_M. For this let $v : M^{\tilde{H}} \to \bigcup_{x\in M} \mathcal{U}(B_x)$ be a unitary which implements β on $M^{\tilde{H}}$. For each $x \in M$ let π_x be in the class of $x \in \Omega = \widehat{B}$ and let $X_x = X/X \cdot \ker \pi_x$ denote the canonical $A_x - B_x$ imprimitivity bimodule. Furthermore, let u^x denote the corresponding action of H_x on X_x. Then (X_x, u^x) becomes a $(\alpha^x, \tau^x) - \beta^x$ imprimitivity pair. For $(x, s) \in M^H$ we define $w(x, s) \in \mathcal{U}(A_x)$ by

$$w(x, s)\xi = u_s^x(\xi)v(x, \tilde{s}), \quad \tilde{s} = sN_\tau.$$

It follows from Proposition 2.1.3 that each $w(x, \cdot)$ is a homomorphism from H_x into $\mathcal{U}(A_x)$ which implements (α^x, τ^x). Thus it remains to show that for each $a \in A_M$ the maps

$$M^H \to E; (x, s) \mapsto w(x, s)a(x) \text{ and } (x, s) \mapsto a(x)w(x, s)$$

are continuous. Using [25, Proposition 1.6 and 1.7], we may realize X as a section space $\Gamma_0(D)$ of a Banach bundle $f : D \to \Omega$ with fibers X^x such that the A- and B-valued inner products and the left A-action and the right B-action on X are given by the inner products and the actions of A_x and B_x on the fibers X_x, respectively (compare with the proof of [20, Lemma 3.11]). It follows from the continuity of the pointwise operations that

$$M^H \to D, (x, s) \mapsto u_s^x(\xi(x))v(x, \tilde{s})$$

is continuous for all $\xi \in X$. Thus, for $a = \langle \xi, \eta \rangle_A$, $\xi, \eta \in X$, we conclude that

$$\begin{aligned} w(x, s)a(x) &= w(x, s)\langle \xi(x), \eta(x) \rangle_{A_x} \\ &= \langle w(x, s)\xi(x), \eta(x) \rangle_{A_x} \\ &= \langle u_s^x(\xi(x))v(x, \tilde{s}), \eta(x) \rangle_{A_x}, \end{aligned}$$

is also continuous in $(x, s) \in M^H$. A similar argument applies for the continuity of the map $(x, s) \to a(x)w(x, s)$.

The fact that $A \rtimes_{\alpha, \tau} \Omega^H$ has continuous trace follows from the same arguments as used at the end of the proof of Proposition 5.2.5. $\square$

PROOF OF THEOREM 5.2.9. We already know from Lemma 5.2.10 that $A \rtimes_{\alpha, \tau} \Omega^H$ has continuous trace, which gives (1) and also implies that $(A \rtimes_{\alpha, \tau} \Omega^H)\hat{}$ is a locally compact Hausdorff space.

In order to prove (2) we construct, as in the proof of [46, Theorem 1.10], an isomorphism $\Psi : A \rtimes_{\alpha, \tau} \Omega^H \to \Gamma_0(\text{res}^* E)$ as follows: Let A be realized as a section algebra $\Gamma_0(E)$ for a C^*-bundle $p : E \to \Omega$ with fibers A_x. For each $x \in \Omega$ let $\rho_x : A \to A_x$ denote evaluation at $x \in \Omega$. Then, if $(x, \rho \times V)$ is an irreducible subgroup representation of $A \rtimes_{\alpha, \tau} \Omega^H$, there exists a unique strictly continuous homomorphism $U : H_x \to \mathcal{U}(A_x)$ such that $(x, \rho \times V)$ is unitarily equivalent to $(x, \rho_x \times U)$ (take $U = \rho^{-1} \circ V$, where ρ is viewed as an isomorphism between A_x and $\mathcal{K}(\mathcal{H}_\rho)$). We define

$$\Psi(f)(x, \rho \times V) = (x, \rho_x \times U)(f).$$

In order to show that Ψ is the desired isomorphism, we start by showing that Ψ is well defined. It follows from [9, Proposition 3.3.7] that $(x, \rho \times V) \mapsto (x, \rho_x \times U)(f)$ vanishes at infinity for all $f \in A \rtimes_{\alpha, \tau} \Omega^H$ (since $\|(x, \rho \times V)(f)\| = \|(x, \rho_x \times$

$U)(f)\|)$. In order to see that this map is also continuous for all $f \in A \rtimes_{\alpha,\tau} \Omega^H$, let us first note that it is enough to show the continuity for the images in $A \rtimes_{\alpha,\tau} \Omega^H$ of all elements $g \otimes a \in \Gamma_c(q^*E)$, $g \in C_c(\Omega^H)$, $a \in A$ ($q : \Omega^H \to \Omega$ denoting the projection) which are defined by

$$g \otimes a(x,s) = g(x,s)a(x),$$

since these elements generate a dense subspace of $A \rtimes_{\alpha,\tau} \Omega^H$ by [56, Lemma 2.1].

So let $g \otimes a$ be as above and let $(x_n, \rho_n \times V^n) \to (x, \rho \times V)$ in $(A \rtimes_{\alpha,\tau} \Omega^H)\widehat{\ }$. By Lemma 5.2.10 we may pass to a subsequence in order to assume that there exists a map $u : M^H \to \bigcup_{y \in M} \mathcal{U}(A_y)$ which implements (α^M, τ^M), where $M = \{x_n; \in \mathbb{N}\} \cup \{x\}$. By Proposition 5.2.2 we know that $(A_M \rtimes_{\alpha,\tau} M^H)\widehat{\ }$ is homeomorphic to $C^*(M^{\tilde{H}})\widehat{\ }$ via the map

$$C^*(M^{\tilde{H}})\widehat{\ } \to (A_M \rtimes_{\alpha,\tau} M^H)\widehat{\ }; (x,\chi) \mapsto (x, \rho_x \times \chi(\rho_x \circ u(x,\cdot))).$$

Therefore we may assume that $(x_n, \chi_n) \to (x,\chi)$ in $C^*(\Omega^{\tilde{H}})\widehat{\ }$ (which contains $C^*(M^{\tilde{H}})\widehat{\ }$) such that $(x_n, \rho_n \times V^n)$ is equivalent to $(x_n, \rho_{x_n} \times \chi_n(\rho_{x_n} \circ u(x_n, \cdot)))$ for all $n \in \mathbb{N}$ and $(x, \rho \times V) = (x, \rho_x \times \chi(\rho_x \circ u(x, \cdot)))$. But then it follows directly from the definitions that

$$(x_n, \rho_{x_n} \times \chi_n(\rho_{x_n} \circ u(x_n, \cdot)))(g \otimes a) = (x_n, \chi_n)(g)\rho_{x_n}(a)$$

$$\to (x,\chi)(g)\rho_x(a) = (x, \rho_x \times \chi(\rho_x \circ u(x, \cdot)))(g \otimes a).$$

Thus, $(x, \rho \times V) \mapsto \Psi(f)(x, \rho \times V)$ is continuous for all $f \in A \rtimes_{\alpha,\tau} \Omega^H$.

It follows directly from the definition of Ψ that it is injective. In order to see that it is onto, it is enough to show that for each $(x, \rho \times V) \in (A \rtimes_{\alpha,\tau} \Omega^H)\widehat{\ }$ evaluation of $\Psi(A \rtimes_{\alpha,\tau} \Omega^H)$ at $(x, \rho \times V)$ has dense image in A_x, and that the image of Ψ is invariant under pointwise multiplication with functions in $C_0((A \rtimes_{\alpha,\tau} \Omega^H)\widehat{\ })$ (this follows by a slight modification of [25, Proposition 1.7]). However, the first assertion is a direct consequence of the definition of Ψ, while the second follows easily from the Dauns-Hoffmann theorem: if $\varphi \in C_c((A \rtimes_{\alpha,\tau} \Omega^H)\widehat{\ })$ and z_φ denotes the element of the center $\mathcal{Z}(A \rtimes_{\alpha,\tau} \Omega^H)$ of $\mathcal{M}(A \rtimes_{\alpha,\tau} \Omega^H)$ which corresponds to φ via the Dauns-Hoffmann theorem (see [50, Corollary 4.4.8]), then it follows from the definition of z_φ that $\Psi(z_\varphi f)(x, \rho \times V) = \varphi(x, \rho \times V)\Psi(f)(x, \rho \times V)$ for all $f \in A \rtimes_{\alpha,\tau} \Omega^H$.

Let us finally assume that $\tilde{G} = G/N_\tau$ is abelian. In order to see that $(A \rtimes_{\alpha,\tau} \Omega^H)\widehat{\ }$ is a σ-proper $\widehat{\tilde{G}}$-space observe that by the continuity and openness of the projection res : $(A \rtimes_{\alpha,\tau} \Omega^H)\widehat{\ } \to \Omega$ we may identify Ω with the quotient space $(A \rtimes_{\alpha,\tau} \Omega^H)\widehat{\ }/\widehat{\tilde{G}}$. Assume that $(x_n)_{n \in \mathbb{N}}$ is a sequence in Ω which converges to some $x \in \Omega$. As above we may pass to a subsequence in order to assume that the action (α^M, τ^M) of M^H on A_M is unitary, where $M = \{x_n; n \in \mathbb{N}\} \cup \{x\}$. It is then a consequence of Proposition 5.2.4 that $q^{-1}(M) = (A_M \rtimes_{\alpha,\tau} M^H)\widehat{\ }$ is a σ-trivial $\widehat{\widetilde{G}}$-space. But using Proposition 5.1.3 this implies that $(A \rtimes_{\alpha,\tau} \Omega^H)\widehat{}$ is a σ-proper $\widehat{\widetilde{G}}$-space. $\square$

Remark 5.2.11. Let (A, G, α, τ) and $H : \Omega \to \mathfrak{K}(G)$ be as in Theorem 5.2.9. Then the inverse of the isomorphism $\Psi : A \rtimes_{\alpha,\tau} \Omega^H \to \Gamma_0(\mathrm{res}^* E)$ is given as follows: Let $\phi : C_0((A \rtimes_{\alpha,\tau} \Omega^H)\widehat{}) \to \mathcal{Z}(A \rtimes_{\alpha,\tau} \Omega^H)$ denote the homomorphism defined by the Dauns-Hofman theorem and let $i_A : A \to \mathcal{M}(A \rtimes_{\alpha,\tau} \Omega^H)$ be as in Lemma 5.2.6. Then define

$$\Phi : C_0((A \rtimes_{\alpha,\tau} \Omega^H)\widehat{}) \otimes_{C(\Omega)} A \to A \rtimes_{\alpha,\tau} \Omega^H; \Phi(\varphi \otimes a) = \phi(\varphi) i_A(a).$$

It is trivially seen that $\Psi \circ \Phi$ is the identity on $C_0((A \rtimes_{\alpha,\tau} \Omega^H)\widehat{}) \otimes_{C(\Omega)} A \cong \Gamma_0(\mathrm{res}^* E)$, where Ψ is defined as in the proof of Theorem 5.2.9 (and then extended to $\mathcal{M}(A \rtimes_{\alpha,\tau} \Omega^H)$). Thus, since Ψ is an isomorphism, it follows that $\Phi = \Psi^{-1}$. Note also that in case where $\widetilde{G} = G/N_\tau$ is abelian it follows easily from the construction of the isomorphism Ψ that it carries the dual action $\widehat{\widetilde{\alpha}}$ of $\widehat{\widetilde{G}}$ on $A \rtimes_{\alpha,\tau} \Omega^H$ to the diagonal action $\theta \otimes_{C(\Omega)} \mathrm{id}$ of $\widehat{\widetilde{G}}$ on $C_0((A \rtimes_{\alpha,\tau} \Omega^H)\widehat{}) \otimes_{C(\Omega)} A$, where θ denotes the natural action of $\widehat{\widetilde{G}}$ on $C_0((A \rtimes_{\alpha,\tau} \Omega^H)\widehat{})$. Thus we obtain complete analogues of [**46**, Theorem 1.10] in case of twisted pointwise unitary subgroup actions.

As a corollary of Theorem 5.1.6 and Theorem 5.2.9 we obtain

COROLLARY 5.2.12. *Suppose that (A, G, α, τ) is a separable twisted covariant system such that A has continuous trace and the action of G on $\Omega = \widehat{A}$ is σ-proper. Suppose further that all stability groups S_x, $x \in \Omega$, are abelian and that the restriction of (α, τ) to Ω^S is pointwise unitary. Then $A \rtimes_{\alpha,\tau} G$ has continuous trace.*

In Section 6.1 we will see that in the situation of Corollary 5.2.12 above, $(A \rtimes_{\alpha,\tau} G)\widehat{}$ is a σ-proper $\widehat{\widetilde{G}}$-space if we assume additionally that $\widetilde{G} = G/N_\tau$ is abelian. This will give a complete answer to the questions raised in the concluding remark on page 297 of [**46**].

5.3. Systems with continuous choices of maximally pointwise unitary subgroups

Let (A, G, α, τ) be a separable abelian twisted system such that A has continuous trace. In this section we give necessary and sufficient conditions for $A \rtimes_{\alpha,\tau} G$ to have continuous trace under the additional assumption that there exists a continuous choice of maximally pointwise unitary subgroups for (A, G, α, τ). We also give a relation between the Dixmier-Douady classes $\delta(A)$ and $\delta(A \rtimes_{\alpha,\tau} G)$ of A and $A \rtimes_{\alpha,\tau} G$, respectively.

Recall that if A is a separable continuous-trace C^*-algebra with $\widehat{A} = \Omega$, then the Dixmier-Douady class $\delta(A)$ of A is a certain element of the third Čech-cohomology group $H^3(\Omega, \mathbb{Z})$ which is a complete invariant for the Ω-stable isomorphism class of A. To be more precise, if A and B are separable continuous trace algebras with $\widehat{A} = \Omega = \widehat{B}$, then $\delta(A) = \delta(B)$ if and only if there exists an isomorphism $\Phi : A \otimes \mathcal{K} \to B \otimes \mathcal{K}$ which induces the identity map on Ω (for more details of this theory see for instance [**9**, Chapter 10] and [**66, 59**]). Since in this memoir we are working with the notion of Morita equivalence rather than stably isomorphism (which in the separable case is the same by [**4**]) we prefer the following formulation (see [**59**, Theorem 3.5]).

PROPOSITION 5.3.1. *Let A and B be two continuous-trace C^*-algebras with paracompact dual spaces $\widehat{A} = \Omega = \widehat{B}$ (here A and B are not assumed to be separable). Then $\delta(A) = \delta(B)$ if and only if A is Morita equivalent to B over Ω, where A is said to be* Morita equivalent to B over Ω *if there exists an $A - B$ imprimitivity bimodule X such that the homeomorphism $\widehat{B} \to \widehat{A}; \pi \to \text{ind}^X \pi$ induces the identity on Ω.*

Note that commutative C^*-algebras always have zero Dixmier-Douady class, so that one might think of the Dixmier-Douady class of A as the obstruction for A being Morita equivalent to a commutative C^*-algebra. If A is a continuous trace algebra with dual space Ω, and if $f : \Lambda \to \Omega$ is a continuous map, then we know from [**55**, Proposition 1.4] that the pull-back f^*A is a continuous-trace C^*-algebra with dual space Λ such that $\delta(f^*A) = f^*(\delta(A))$, where $f^* : H^3(\Omega, \mathbb{Z}) \to H^3(\Lambda, \mathbb{Z})$ denotes the homomorphism induced by f. We use this result for the proof of

THEOREM 5.3.2. *Suppose that (A, G, α, τ) is a separable twisted abelian system such that A has continuous trace. Assume that $H : \Omega = \widehat{A} \to \mathfrak{K}(G); x \mapsto H_x$ is a continuous choice of maximally pointwise unitary subgroups of G for (A, G, α, τ). Then $A \rtimes_{\alpha,\tau} G$ has continuous trace if and only if the canonical action of G on $(A \rtimes_{\alpha,\tau} \Omega^H)\widehat{\ }$ is σ-proper. Moreover, if this condition is satisfied and if*

$$\text{res} : (A \rtimes_{\alpha,\tau} \Omega)\widehat{\ } \to \Omega; \text{res}(x, \rho \times V) = x$$

denotes the restriction map (via the identification of Ω with $\widehat{A}$) and

$$\text{ind} : (A \rtimes_{\alpha,\tau} \Omega^H)\widehat{\ } \to (A \rtimes_{\alpha,\tau} G)\widehat{\ }; \text{ind}(x, \rho \times V) = \text{ind}_{H_x}^G(\rho \times V)$$

denotes the induction map (which is well defined and surjective by Theorem 3.2.1) then

$$\text{res}^*(\delta(A)) = \text{ind}^*(\delta(A \rtimes_{\alpha,\tau} G)).$$

The relation between the Dixmier-Douady classes of A and $A \rtimes_{\alpha,\tau} G$ given above generalizes similar relations given in [**54**, Corollary 2.5] for actions which are pointwise unitary on the constant stabilizer and [**19**, Theorem 6] for systems with constant Mackey obstructions. Note that it was already pointed out by

Raeburn and Rosenberg that this relation between $\delta(A)$ and $\delta(A \rtimes_{\alpha,\tau} G)$ does not determine $\delta(A \rtimes_{\alpha,\tau} G)$ in general. If this were the case, then $\delta(C_0(\Omega) \rtimes G)$ would be trivial for all abelian transformation group algebras with continuous trace, since the Dixmier-Douady class of a commutative C^*-algebra is always zero. But Raeburn and Rosenberg constructed an action of $\mathbb{R}$ on S^3 with constant stabilizer $\mathbb{Z}$ such that $C(S^3) \rtimes \mathbb{R}$ has continuous trace but $\delta(C(S^3) \rtimes \mathbb{R}) \neq 0$ (see [**54**, Example 4.6]).

PROOF OF THEOREM 5.3.2. Using Proposition 4.4.6 and the results presented in Chapter 2 (especially Propositions 2.1.4 and 2.2.1) we may pass to the double dual action $\widehat{\widehat{\alpha}}$ in order to assume that G is abelian and τ is trivial.

Since the restriction of α to Ω^H is pointwise unitary, we conclude from Theorem 5.2.9 that $A \rtimes_\alpha \Omega^H$ has continuous trace. Let γ^H denote the canonical action of G on $A \rtimes_\alpha \Omega^H$. Then it follows from Lemma 3.1.10 that for each $(x, \rho \times V) \in (A \rtimes_{\alpha,\tau} \Omega^H)\widehat{}$ its stabilizer $S_{(x,\rho \times V)}$, under the corresponding action of G on $(A \rtimes_{\alpha,\tau} \Omega^H)\widehat{}$, is equal to H_x. Thus we conclude that the stabilizer map $S : (A \rtimes_{\alpha,\tau} \Omega^H)\widehat{} \to \mathfrak{K}(G)$ for the system $(A \rtimes_{\alpha,\tau} \Omega^H, G, \gamma^H)$ is continuous. Moreover, it is easily verified that $(x, \rho \times V) \times V$ is a covariant representation of $(A \rtimes_{\alpha,\tau} \Omega^H, S_{(x,\rho \times V)}, \gamma^H)$ for each $(x, \rho \times V) \in (A \rtimes_{\alpha,\tau} \Omega^H)\widehat{}$. Thus, if we put $\Lambda = (A \rtimes_{\alpha,\tau} \Omega^H)\widehat{}$, then γ^H restricts to a pointwise unitary action of Λ^S on $A \rtimes_\alpha \Omega^H$. Since G is abelian, it follows now from Corollary 5.2.12 that $(A \rtimes_\alpha \Omega^H) \rtimes_{\gamma^H} G$ has continuous trace, whenever G acts σ-properly on $(A \rtimes_{\alpha,\tau} \Omega^H)\widehat{}$. But it is a consequence of Proposition 4.3.5 (applied to the regular pair (id, H), where $\mathrm{id} : \widehat{A} = \Omega \to \Omega$ denotes the identity) that $A \rtimes_{\alpha,\tau} G$ is a quotient of $(A \rtimes_\alpha \Omega^H) \rtimes_{\gamma^H} G$, which implies that $A \rtimes_\alpha G$ has continuous trace if G acts σ-properly on $(A \rtimes_{\alpha,\tau} \Omega^H)\widehat{}$.

For the opposite direction assume that $A \rtimes_\alpha G$ has continuous trace. Let $\mathrm{Res}^G : (A \rtimes_\alpha G)\widehat{} \to \mathcal{Q}(\widehat{A})$ denote the quasi-orbit map, i.e., Res^G maps $\pi \times U \in (A \rtimes_\alpha G)\widehat{}$ to the unique quasi-orbit $\mathcal{Q}(\rho) \subseteq \widehat{A}$ satisfying $\ker \pi = \bigcap_{s \in G} \ker(\rho \circ \alpha_s)$. By Theorem 3.3.3 we know that $H^\perp : (A \rtimes_\alpha G)\widehat{} \to \mathfrak{K}(\widehat{G})$ defined by $H^\perp_{\pi \times U} = (H_{\mathrm{Res}^G(\pi \times U)})^\perp$ is a choice of maximally pointwise unitary subgroups for $(A \rtimes_\alpha G, \widehat{G}, \widehat{\alpha})$, which is continuous since Res^G and H are continuous (note that H is constant on quasi-orbits by definition). Thus, if we put $\widehat{\Omega} = (A \rtimes_\alpha G)\widehat{}$, it follows from Proposition 5.2.9 that $(A \rtimes_\alpha G) \rtimes_{\widehat{\alpha}} \widehat{\Omega}^{H^\perp}$ has continuous trace and the action $\widehat{\widehat{\alpha}}$ of G on $((A \rtimes_\alpha G) \rtimes_{\widehat{\alpha}} \widehat{\Omega}^{H^\perp})\widehat{}$ is σ-proper.

Now let $\widetilde{\Omega}$ denote the closure of the image of H in $\mathfrak{K}(G)$. Let $R : \widehat{A} \to \widetilde{\Omega}$ be defined by $R(\rho) = H_\rho$ and let $L : \widetilde{\Omega} \to \mathfrak{K}(G)$ denote the identity. Then (R, L) is a pair of regular maps for (A, G, α) with base $\widetilde{\Omega}$ such that $L \circ R = H \circ \mathrm{id}$, where again $\mathrm{id} : \widehat{A} = \Omega \to \Omega$ denotes the identity. We conclude from Proposition 4.1.7 that $A \rtimes_\alpha \Omega^H$ is G-isomorphic to $C^*(\mathcal{S}_R^L)$. Now let $(\widehat{R}, L^\perp)$ denote the pair of regular maps for $(A \rtimes_\alpha G, \widehat{G}, \widehat{\alpha})$ which is dual to (R, L) (see Definition 4.4.1). Then it follows directly from the definitions that $L^\perp \circ \widehat{R} = H^\perp \circ \mathrm{id}$. Thus, Proposition 4.1.7 implies that $(A \rtimes_\alpha G) \rtimes_{\widehat{\alpha}} \widehat{\Omega}^{H^\perp}$ is isomorphic to $C^*(\mathcal{S}_{\widehat{R}}^{L^\perp})$, and it follows easily from the construction of this isomorphism that it is equivariant with respect to the double dual actions $\widehat{\widehat{\alpha}}$ of G on both algebras. But by Theorem 4.4.2 we know that the double dual action $\widehat{\widehat{\alpha}}$ of G on $C^*(\mathcal{S}_R^{L^\perp})$ is Morita equivalent to the canonical action γ^L of G on $C^*(\mathcal{S}_R^L)$. Putting things together, we conclude that the double dual action $\widehat{\widehat{\alpha}}$ of G on $(A \rtimes_\alpha G) \rtimes_{\widehat{\alpha}} \widehat{\Omega}^{H^\perp}$ is Morita equivalent to the canonical action γ^H of G on $A \rtimes_\alpha \Omega^H$. Since we already know that the corresponding action of G on $((A \rtimes_\alpha G) \rtimes_{\widehat{\alpha}} \widehat{\Omega}^{H^\perp})^\widehat{}$ is σ-proper, this implies that G acts σ-properly on $(A \rtimes_\alpha \Omega^H)^\widehat{}$, too.

In order to complete the proof assume that G acts σ-properly on $(A \rtimes_\alpha \Omega^H)^\widehat{}$. By Theorem 5.2.9 we know that $A \rtimes_\alpha \Omega^H \cong \text{res}^* A$ and that $(A \rtimes_\alpha G) \rtimes_{\widehat{\alpha}} \widehat{\Omega}^{H^\perp} \cong \widetilde{\text{res}}^*(A \rtimes_\alpha G)$, where $\widetilde{\text{res}} : ((A \rtimes_\alpha G) \rtimes_{\widehat{\alpha}} \widehat{\Omega}^{H^\perp})^\widehat{} \to (A \rtimes_\alpha G)^\widehat{}$ denotes restriction. Thus the theorem follows from Proposition 5.3.1 and the remarks following it, if the homeomorphism between $(A \rtimes_\alpha \Omega^H)^\widehat{}$ and $((A \rtimes_\alpha G) \rtimes_{\widehat{\alpha}} \widehat{\Omega}^{H^\perp})^\widehat{}$, given by the Morita equivalence constructed above, intertwines ind with $\widetilde{\text{res}}$. But this follows from Theorem 2.2.2 together with a careful look at the construction of the imprimitivity bimodule given in the proof of Theorem 4.4.2. $\square$

If all Mackey obstructions of (A, G, α, τ) vanish (which is always true if $\tilde{G} = G/N_\tau$ is isomorphic to $\mathbb{R}$, $\mathbb{Z}$ or $\mathbb{T}$), then the maximally pointwise unitary subgroups of G are just the stabilizers of the elements of $\widehat{A}$. As a corollary of Theorem 5.3.2 we get

THEOREM 5.3.3. *Let (A, G, α, τ) be a separable twisted abelian system such that A has continuous trace and all Mackey obstructions of (A, G, α, τ) are trivial. Then $A \rtimes_{\alpha,\tau} G$ has continuous trace if and only if the stabilizer map $S : \Omega = \widehat{A} \to \mathfrak{K}(G)$ is continuous and the canonical action of G on $(A \rtimes_{\alpha,\tau} \Omega^S)^\widehat{}$ is σ-proper. Moreover, if this is true then*

$$\text{res}^*(\delta(A)) = \text{ind}^*(\delta(A \rtimes_{\alpha,\tau} G)),$$

where $\text{res} : (A \rtimes_{\alpha,\tau} \Omega^S)^\widehat{} \to \Omega, \text{res}(x, \rho \times V) = x$ denotes the restriction map and $\text{ind} : (A \rtimes_{\alpha,\tau} \Omega^S)^\widehat{} \to (A \rtimes_{\alpha,\tau} G)^\widehat{}; \text{ind}(x, \rho \times V) = \text{ind}_{S_x}^G(\rho \times V)$ denotes the induction map.

PROOF. Since all Mackey obstructions are assumed to be trivial, the theorem follows from Theorem 5.3.2 as soon as we have shown that the continuity of S is necessary for $A \rtimes_{\alpha,\tau} G$ to have continuous trace.

So assume that $A \rtimes_{\alpha,\tau} G$ has continuous trace and let $\rho_n \to \rho$ in $\widehat{A}$. Since $\mathfrak{K}(G)$ is compact and $\widehat{A}$ is Hausdorff we may assume, by passing to a subsequence, that $S_{\rho_n} \to S_0$ in $\mathfrak{K}(G)$ for some $S_0 \subseteq S_\rho$. By the openness of the quasi-orbit map $\text{Res}^G : (A \rtimes_{\alpha,\tau} G)^\widehat{} \to \mathcal{Q}(\widehat{A})$ we find, again by passing to a subsequence if necessary, a sequence $(\pi_n \times U^n)_{n\in\mathbb{N}} \subseteq (A \rtimes_{\alpha,\tau} G)^\widehat{}$ converging to some $\pi \times U \in (A \rtimes_{\alpha,\tau} G)^\widehat{}$ such that $\rho \in \text{Res}^G(\pi \times U)$ and $\rho_n \in \text{Res}^G(\pi_n \times U^n)$ for all $n \in \mathbb{N}$. By Theorem 3.3.3 we know that the stabilizer of $\pi_n \times U^n$ in $\widehat{\tilde{G}}$ is equal to $S_{\rho_n}^\perp$ since by assumption $\Sigma_{\rho'} = S_{\rho'}$ for all $\rho' \in \widehat{A}$. Since $(A \rtimes_{\alpha,\tau} G)^\widehat{}$ is Hausdorff, it follows that the stabilizer of $\pi \times U$, which is just $S_\rho^\perp$, is contained in $S_0^\perp = \lim_{n \to \infty} S_{\rho_n}^\perp$. But this implies that $S_0 = S_\rho$. Hence, S is continuous. $\square$

5.4. Continuous trace for systems with Hausdorff quasi-orbit space

We will now see that for $A \rtimes_{\alpha,\tau} G$ to have continuous trace, it is necessary that the maximally pointwise unitary subgroups vary continuously — at least in the weak sense of the following definition.

DEFINITION 5.4.1. Suppose that (A, G, α, τ) is a separable twisted abelian system such that A has continuous trace. We say that (A, G, α, τ) has *continuously varying maximally pointwise unitary subgroups* if the following is true: If $\rho_n \to \rho$ in $\widehat{A}$ and if H_{ρ_n} is a maximally ρ_n-unitary subgroup of G for all $n \in \mathbb{N}$ such that $H_{\rho_n} \to H$ for some $H \in \mathfrak{K}(G)$. Then H is a maximally ρ-unitary subgroup of G.

Note that the condition that (A, G, α, τ) has continuously varying maximally pointwise unitary subgroups does not imply that there exists a continuous choice of maximally pointwise unitary subgroups for (A, G, α, τ) (see Example 6.3.1 below). However, at least in case where the quasi-orbit space $\mathcal{Q}(\widehat{A})$ of $\widehat{A}$ is Hausdorff we will see later that there are continuous subchoices on closed G-invariant subsets of $\widehat{A}$, as defined below, which are big enough for our purposes.

DEFINITION 5.4.2. Suppose that (A, G, α, τ) is a separable twisted abelian system such that A has continuous trace and let M be a closed G-invariant subset of $\widehat{A}$. If $H : M \to \mathfrak{K}(G)$ is a continuous choice of maximally pointwise unitary subgroups for (A_M, G, α, τ), then H is called a *continuous subchoice of maximally pointwise unitary subgroups* for (A, G, α, τ).

We are now prepared to formulate

THEOREM 5.4.3. *Suppose that (A, G, α, τ) is a separable twisted abelian system such that A has continuous trace and such that the quasi-orbit space $\mathcal{Q}_G(\widehat{A})$ is Hausdorff. Then $A \rtimes_{\alpha,\tau} G$ has continuous trace if and only if the following conditions are satisfied*

(1) *(A, G, α, τ) has continuously varying maximally pointwise unitary subgroups.*

(2) *If $H : M \to \mathfrak{K}(G)$ is any continuous subchoice of maximally pointwise unitary subgroups for (A, G, α, τ), then the canonical action of G on $(A_M \rtimes_{\alpha,\tau} M^H)\widehat{\ }$ is σ-proper.*

Moreover, if $\mathcal{Q}_G(\widehat{A})$ is not assumed to be Hausdorff, then Conditions (1) and (2) above are (at least) necessary for $A \rtimes_{\alpha,\tau} G$ to have continuous trace.

Note that we do not know whether Theorem 5.4.3 remains true without the Hausdorff assumption on the quasi-orbit space. The reason for this assumption is that it guarantees the existence of continuous subchoices of maximally pointwise unitary subgroups on closed subsets of $\widehat{A}$ which are big enough to imply that $A \rtimes_{\alpha,\tau} G$ has continuous trace. Let us note that the Hausdorff assumption on the quasi-orbit space is not necessary for $A \rtimes_{\alpha,\tau} G$ to have continuous trace, as is shown in Example 6.3.1 below.

For the proof of the theorem we need a kind of semi-continuity result for the maximally pointwise unitary subgroups of a separable twisted abelian system (A, G, α, τ). In order to obtain such a result we need a proposition which shows that the Mackey obstructions of a given twisted system (A, G, α, τ) vary continuously in a certain sense. We are very grateful to Dana Williams for providing this result and for allowing us to use the material in this work. Note that the following proposition will also play an important role in the classification of crossed products with continuous trace by systems with continuously varying stabilizers as given in the next section.

PROPOSITION 5.4.4 (WILLIAMS). *Suppose that* (A, G, α, τ) *is a separable twisted covariant system such that A has continuous trace with dual space* $\widehat{A} = \Omega$. *Let $x_0 \in \Omega$ and let $s_0, t_0 \in S_{x_0}$. Then there exists a neighborhood U of x_0 and a choice of representatives $\omega_x \in Z^2(\tilde{S}_x, \mathbb{T})$ for the Mackey obstructions of $x \in U$ such that, viewed as a function on $\{(x, s, t) \in U \times G \times G; s, t \in S_x\}$, $\omega_x(s, t)$ is continuous near both (x_0, s_0, t_0) and (x_0, t_0, s_0).*

PROOF. Since the Mackey obstructions are invariant under passing over to Morita equivalent actions we may assume that τ is trivial and A is stable. We then find an open neighborhood W of x_0 such that $\overline{W}$ is compact and $A_{\overline{W}} \cong C(\overline{W}, \mathcal{K})$. We may identify the ideal A_W of A with the ideal $C_0(W, \mathcal{K}) \subseteq C(\overline{W}, \mathcal{K})$. By the continuity of the G-action on Ω we can find neighborhoods $U \subseteq W$ of x_0 and V_1, V_2, V_3 and V_4 of $s_0, t_0, s_0 t_0$ and $t_0 s_0$, respectively, so that, if $x \in U$ and $s \in V = V_1 \cup V_2 \cup V_3 \cup V_4$, then $sx \in W$. Let

$$\mathcal{P}' = \{(x, s) \in W \times G; s \in S_x\}.$$

Identifying $\mathrm{Aut}(\mathcal{K})$ with the projective unitary group $\mathcal{PU}$, we obtain a map $q: \mathcal{P}' \to \mathcal{PU}$ as follows: For each $x \in W$ let $\Psi^x: A \to \mathcal{K}$ be the homomorphism defined by composing the quotient map $A \to A_{\overline{W}} = C(\overline{W}, \mathcal{K})$ with evaluation at x. Then Ψ^x transports the restriction of α to S_x to an action of S_x on $\mathcal{K}$, which defines a continuous homomorphism $q(x, \cdot): S_x \to \mathcal{PU}$ via the canonical identification of $\mathcal{PU}$ with $\mathrm{Aut}(\mathcal{K})$.

Observe that if γ' is any Borel cross section for the quotient map $p: \mathcal{U} \to \mathcal{PU}$, and if $u'(x, s) = \gamma'(q(x, s))$, then we obtain cocycles $\omega'_x \in Z^2(S_x, \mathbb{T})$ representing the Mackey obstruction of $x \in W$ by

$$\omega'_x(s, t) = u'(x, s)u'(x, t)u'(x, st)^{-1}.$$

Let $f \in C_0(U, \mathcal{K}) \subseteq C_0(W, \mathcal{K}) = A_W \subseteq A$. Since $sU \subseteq W$ for all $s \in V$ we observe that $\alpha_s(f) \in C_0(W, \mathcal{K})$ for all $s \in V$. Define

$$\mathcal{P} = \{(x, s) \in \mathcal{P}'; x \in U \text{ and } s \in V\}.$$

Then it follows from the construction of q that $q(x,s)f(x)q(x,s)^* = \alpha_s(f)(x)$ for all $(x,s) \in \mathcal{P}$ and $f \in C_0(U,\mathcal{K})$. Using this observation, we claim that $q|_\mathcal{P}$ is continuous. For this suppose that $(x_n, s_n) \to (x,s) \in \mathcal{P}$, and let $T \in \mathcal{K}$. Choose $f \in C_0(U,\mathcal{K})$ such that $f(x) = T$ in a neighborhood of x. Since $\alpha_{s_n}(f)(x_n)$ converges to $\alpha_s(f)(x)$, we have $\text{Ad}\, q(x_n, s_n)(T) \to \text{Ad}\, q(x,s)(T)$ for all $T \in \mathcal{K}$, which implies that $q(x_n, s_n) \to q(x,s)$ in $\mathcal{P}\mathcal{U}$.

Now let γ' be a Borel cross section for the quotient map $p : \mathcal{U} \to \mathcal{P}\mathcal{U}$. Since $\ker p = \mathbb{T}\,\text{Id}$, it follows from [**26**] that there are local continuous sections for $p : \mathcal{U} \to \mathcal{P}\mathcal{U}$. Thus we may alter γ' to obtain a Borel cross section γ for p which is continuous near $q(x_0, s_0)$, $q(x_0, t_0)$, $q(x_0, s_0 t_0)$ (and $q(x_0, t_0 s_0)$ if G is not abelian). Define $u(x,s) = \gamma(q(x,s))$ and let

$$\omega_x(s,t) = u(x,s)u(x,t)u(x,st)^{-1}.$$

Then $(x,s,t) \mapsto \omega_x(s,t)$ is continuous near (x_0, s_0, t_0) and (x_0, t_0, s_0) since u is continuous near $(x_0, s_0), (x_0, t_0), (x_0, s_0 t_0)$ and $(x_0, t_0 s_0)$. $\square$

The following corollary is what we need for the proof of Theorem 5.4.3.

COROLLARY 5.4.5. *Let (A, G, α, τ) be a separable twisted abelian system such that A has continuous trace. Assume that $(\rho_n)_{n \in \mathbb{N}}$ is a sequence in $\widehat{A}$ converging to some $\rho_0 \in \widehat{A}$. For each $n \in \mathbb{N}$ let H_n be a maximally ρ_n unitary subgroup of G such that $H_n \to H_0$ for some $H_0 \in \mathfrak{K}(G)$. Then H_0 is contained in a maximally ρ-unitary subgroup of G.*

PROOF. Let $\omega_\rho \in Z^2(\tilde{S}_{\rho_0}, \mathbb{T})$ be in the class of the Mackey obstruction of ρ. We have to show that the restriction of ω_ρ to H_0 is trivial. For this it is enough to show that $\omega_{\rho_0}(s,t) = \omega_{\rho_0}(t,s)$ for all $s, t \in H_0$. Let $s_0, t_0 \in H_0$. By Williams's proposition we may assume that there is a neighborhood U of $\rho_0 \in \widehat{A}$ and representatives $\omega_\rho \in Z^2(\tilde{S}_\rho, \mathbb{T})$ such that $(\rho, s, t) \to \omega_\rho(s,t)$ is continuous near (ρ_0, s_0, t_0) and (ρ_0, t_0, s_0). Since H_0 is the limit of the H_n we may assume that there exist elements $s_n, t_n \in H_n$ such that $s_n \to s_0$ and $t_n \to t_0$. Thus we obtain

$$\omega_{\rho_0}(s_0, t_0) = \lim_{n \to \infty} \omega_{\rho_n}(s_n, t_n) = \lim_{n \to \infty} \omega_{\rho_n}(t_n, s_n) = \omega_{\rho_0}(t_0, s_0).$$

$\square$

The following corollary of Proposition 5.4.4, which will be needed in the next section, is also due to Dana Williams. Note that in the proof of the corollary above we used the same arguments as were used by Williams in the proof of the corollary below.

COROLLARY 5.4.6 (WILLIAMS). *Suppose that (A, G, α, τ) is a separable twisted abelian system such that the stabilizer map $S : \widehat{A} \to \mathfrak{K}(G)$ is continuous. Suppose further that $\rho_n \to \rho_0$ in $\widehat{A}$ such that the symmetrizers Σ_{ρ_n} converge to a group Σ in $\mathfrak{K}(G)$. Then Σ is contained in the symmetrizer Σ_{ρ_0} of ρ_0.*

PROOF. Let $\omega_{\rho_0} \in Z^2(\tilde{S}_{\rho_0}, \mathbb{T})$ be in the class of the Mackey obstruction of ρ and let $s_0 \in \Sigma$ and $t_0 \in S_{\rho_0}$. We have to show that $\omega_{\rho_0}(s_0, t_0) = \omega_{\rho_0}(t_0, s_0)$. By Proposition 5.4.4 we may assume that there is a neighborhood U of $\rho_0 \in \widehat{A}$ and representatives $\omega_\rho \in Z^2(\tilde{S}_\rho, \mathbb{T})$ such that $(\rho, s, t) \to \omega_\rho(s, t)$ is continuous near (ρ_0, s_0, t_0) and (ρ_0, t_0, s_0). Since Σ is the limit of the Σ_{ρ_n} we may pass to a subsequence in order to assume that there exist elements $s_n \in \Sigma_{\rho_n}$ such that $s_n \to s_0$. Similarly, we find a sequence $(t_n)_{n \in \mathbb{N}}$ such that $t_n \in S_{\rho_n}$ for all $n \in \mathbb{N}$ and $t_n \to t_0$. Thus we obtain

$$\omega_{\rho_0}(s_0, t_0) = \lim_{n \to \infty} \omega_{\rho_n}(s_n, t_n) = \lim_{n \to \infty} \omega_{\rho_n}(t_n, s_n) = \omega_{\rho_0}(t_0, s_0).$$

$\square$

PROOF OF THEOREM 5.4.3. As usual we pass to a Morita equivalent action in order to assume that τ is trivial. So let (A, G, α) be a separable abelian system such that A and $A \rtimes_\alpha G$ have continuous trace. Let $\rho_n \to \rho$ in $\widehat{A}$ and assume that for each $n \in \mathbb{N}$, H_n is a maximally ρ_n-unitary subgroup of G such that $H_n \to H_0$ in $\mathfrak{K}(G)$. By Corollary 5.4.5 we conclude that H_0 is contained in a maximally ρ-unitary subgroup, say H, of G. Let us assume that $H \neq H_0$ and let $\text{Res}^G : (A \rtimes_\alpha G)^\wedge \to \mathcal{Q}_G(\widehat{A})$ denote the canonical map (see Proposition 3.3.1). For each ρ_n let $\pi_n \times U^n \in (A \rtimes_\alpha G)^\wedge$ such that $\rho_n \in \text{Res}^G(\pi_n \times U^n)$, and let $\pi \times U \in (A \rtimes_\alpha G)^\wedge$ such that $\rho \in \text{Res}^G(\pi \times U)$. Since Res^G is open we may pass to a subsequence and assume that $\pi_n \times U^n \to \pi \times U$. By Theorem 3.3.3 we know that $H_n^\perp$ is a maximally $\pi_n \times U^n$-unitary subgroup of $\widehat{G}$ and that $H^\perp$ is maximally $\pi \times U$-unitary. But by the continuity of $L \mapsto L^\perp$ it follows that $H_n^\perp \to H_0^\perp$ which properly contains $H^\perp$. But this contradicts Corollary 5.4.5 which says that $H_0^\perp$ should be contained in a maximally $\pi \times U$-unitary subgroup of $\widehat{G}$. Thus, $H = H_0$ and (A, G, α) satisfies Condition (1) of the theorem (compare with the proof of Theorem 5.3.3).

Condition (2) follows directly from Theorem 5.3.2 and the fact that if M is any closed G-invariant subset of $\widehat{A}$, then A_M and $A_M \rtimes_\alpha G$ have continuous trace by assumption. Note that we haven't used the Hausdorff assumption on $\mathcal{Q}_G(\widehat{A})$ so far.

Suppose now that (A, G, α) satisfies Conditions (1) and (2) of the theorem and that $\mathcal{Q}_G(\widehat{A})$ is Hausdorff. Let $\pi_n \times U_n \to \pi \times U$ in $(A \rtimes_\alpha G)^\wedge$. By Proposition 5.1.4 we have to show that $(\pi_n \times U^n)_{n \in \mathbb{N}}$ has a subsequence whose closure is contained in a continuous trace quotient of $A \rtimes_\alpha G$. If there is a subsequence of $(\pi_n \times U^n)_{n \in \mathbb{N}}$ such that Res^G is constant on this subsequence, then it is easily seen that the closure of this subsequence is contained in $(A_M \rtimes_{\alpha, \tau} G)^\wedge$, for $M = q^{-1}(\text{Res}^G(\pi \times U))$, $q : \widehat{A} \to \mathcal{Q}_G(\widehat{A})$ denoting the quotient map. Since maximally pointwise unitary subgroups may be chosen to be constant on quasi-orbits, we conclude that there exists a continuous (in this case even constant) choice $H : M \to \mathfrak{K}(G)$ of maximally pointwise unitary subgroups on the closed subset M of $\widehat{A}$. Thus (2) implies that the canonical action of G on $(A \rtimes_\alpha M^H)^\wedge$

is σ-proper, from which follows by Theorem 5.3.2 that $A_M \rtimes_\alpha G$ has continuous trace.

If there is no such subsequence of $(\pi_n \times U^n)_{n \in \mathbb{N}}$, then we may assume (again by passing to a subsequence) that $\operatorname{Res}^G(\pi_n \times U^n) \neq \operatorname{Res}^G(\pi_m \times U^m) \neq \operatorname{Res}^G(\pi \times U)$ for all $n, m \in \mathbb{N}$ such that $n \neq m$. Let $\rho_n \to \rho$ in $\widehat{A}$ such that $\rho_n \in q^{-1}(\operatorname{Res}^G(\pi_n \times U^n))$ and $\rho \in q^{-1}(\operatorname{Res}^G(\pi \times U))$. The existence of such a sequence follows from the openness of q by passing to another subsequence. For each $n \in \mathbb{N}$ let H_n be maximally ρ_n-unitary. Since $\mathfrak{K}(G)$ is compact we may assume that $H_n \to H_0$ for some $H_0 \in \mathfrak{K}(G)$ which is maximally ρ-unitary by (1). Now let $\tilde{M} = \{\operatorname{Res}^G(\pi_n \times U^n); n \in \mathbb{N}\} \cup \{\operatorname{Res}^G(\pi \times U)\}$. Then $\tilde{M}$ is closed in $\mathcal{Q}_G(\widehat{A})$ by the Hausdorff assumption on $\mathcal{Q}_G(\widehat{A})$. Thus $M = q^{-1}(\tilde{M})$ is closed in $\widehat{A}$, and we may define a continuous subchoice of maximally pointwise unitary subgroups $H : M \to \mathfrak{K}(G)$ by defining $H_\sigma = H_n$ if $\sigma \in q^{-1}(\operatorname{Res}^G(\pi_n \times U^n))$ and $H_\sigma = H_0$ if $\sigma \in q^{-1}(\operatorname{Res}^G(\pi \times U))$. It is clear that the closure of $(\pi_n \times U^n)_{n \in \mathbb{N}}$ is contained in $(A_M \rtimes_\alpha G)\widehat{}$, and it is again a consequence of (2) and Theorem 5.3.2 that $A_M \rtimes_\alpha G$ has continuous trace. The theorem follows now from Proposition 5.1.4. $\square$

Note that the proof fails in case where $\mathcal{Q}_G(\widehat{A})$ is not assumed to be Hausdorff since in this situation the set $\tilde{M}$ as defined above is not closed in general, and it is not clear whether the above defined map $H : q^{-1}(\tilde{M}) \to \mathfrak{K}(G)$ can be extended continuously to the closure. We will now see that in case where G/N_τ is compact or discrete no such problems appear.

COROLLARY 5.4.7. *Let (A, G, α, τ) be a separable abelian twisted system such that A has continuous trace and such that $\tilde{G} = G/N_\tau$ is compact. Then $A \rtimes_{\alpha,\tau} G$ has continuous trace if and only if (A, G, α, τ) has continuously varying maximally pointwise unitary subgroups.*

PROOF. If $\tilde{G}$ is compact then $\mathcal{Q}_G(\widehat{A}) = \widehat{A}/G$ is automatically Hausdorff. Thus the proof follows from Theorem 5.4.3 and the fact that an action of a compact group on a Hausdorff locally compact space is σ-proper if and only if the stabilizer map is continuous. $\square$

COROLLARY 5.4.8. *Let (A, G, α, τ) be a separable abelian twisted system such that A has continuous trace and such that $\tilde{G} = G/N_\tau$ is discrete. Then $A \rtimes_{\alpha,\tau} G$ has continuous trace if and only if the following conditions are satisfied:*
   (1) *The quasi-orbit space $\mathcal{Q}_G(\widehat{A})$ is Hausdorff.*
   (2) *(A, G, α, τ) has continuously varying maximally pointwise unitary subgroups.*
   (3) *If $H : M \to \mathfrak{K}(G)$ is any continuous subchoice of maximally pointwise unitary subgroups for (A, G, α, τ), then the canonical action of G on $(A_M \rtimes_{\alpha,\tau} M^H)\widehat{}$ is σ-proper.*

PROOF. The proof follows from Theorem 5.4.3 as soon as we have shown that if $A \rtimes_{\alpha,\tau} G$ has continuous trace, then $\mathcal{Q}_G(\widehat{A})$ has to be Hausdorff. But by

Proposition 3.3.1 we know that $\mathcal{Q}_G(\widehat{A})$ is homeomorphic to $\mathcal{Q}_{\widehat{G}}((A \rtimes_{\alpha,\tau} G)\widehat{\ })$, which is Hausdorff if $(A \rtimes_{\alpha,\tau} G)\widehat{\ }$ is Hausdorff since $\widehat{G}$ is compact. □

As mentioned before, Example 6.3.1 shows that the Hausdorff assumption on $\mathcal{Q}_G(\widehat{A})$ is in general not necessary for $A \rtimes_{\alpha,\tau} G$ to have continuous trace.

5.5. Systems with continuously varying stabilizers

In this section we want to use the results of the previous sections in order to describe crossed products with continuous trace of abelian twisted systems (A, G, α, τ) under the additional assumption that the stabilizer map $S : \Omega = \widehat{A} \to \mathfrak{K}(G); x \mapsto S_x$ is continuous. Before we state the main result recall (for instance from [19]) that in the most elementary case of an action α of the abelian group G on $\mathcal{K}$, $\mathcal{K} \rtimes_{\alpha} G$ has continuous trace if and only if the Mackey obstruction $[\omega]$ of the system is type I. Baggett's and Kleppner's result implies that this is true if and only if the canonical homomorphism $h_{\omega} : G \to \widehat{G}$ has closed range and is open as a map onto its image. Suppose now that (A, G, α) is a separable abelian system such that A has continuous trace and such that the action of G on $\Omega = \widehat{A}$ is trivial. Then each $x \in \Omega$ defines a canonical map $h_{\omega_x} : G \to \widehat{G}$ via the Mackey obstruction $[\omega_x] \in H^2(G, \mathbb{T})$ of $x \in \Omega$. Thus we obtain a canonical map

$$h^{\alpha} : \Omega \times G \to \Omega \times \widehat{G}; (x, s) \mapsto (x, h_{\omega_x}(s)).$$

We will see that, in complete analogy to the case of actions on $\mathcal{K}$, the crossed product $A \rtimes_{\alpha} G$ has continuous trace if and only if h^{α} has closed range and is open as a map onto its image. In fact we will investigate the more general case of twisted abelian systems (A, G, α, τ) with continuously varying stabilizers. In this situation each $x \in \Omega = \widehat{A}$ defines a homomorphism $h_{\omega_x} : \tilde{S}_x \to \widehat{\tilde{S}}_x$, for $\tilde{S}_x = S_x/N_\tau$, which is defined by the Mackey obstruction $[\omega_x] \in H^2(\tilde{S}_x, \mathbb{T})$ at x. As above, these maps can be put together in order to obtain a map

$$h^{\alpha,\tau} : \Omega^{\tilde{S}} \to C^*(\Omega^{\tilde{S}})\widehat{\ } \cong \Omega \times_{S^\perp} \widehat{\widehat{G}}; (x, \tilde{s}) \mapsto (x, h_{\omega_x}(\tilde{s})).$$

DEFINITION 5.5.1. Let (A, G, α, τ) be a separable abelian twisted system such that A has continuous trace and the stabilizer map $S : \Omega = \widehat{A} \to \mathfrak{K}(G)$ is continuous. The map

$$h^{\alpha,\tau} : \Omega^{\tilde{S}} \to \Omega \times_{S^\perp} \widehat{\widehat{G}}; h^{\alpha,\tau}(x, \tilde{s}) = (x, h_{\omega_x}(\tilde{s}))$$

is called the *Mackey obstruction map* of (A, G, α, τ).

Since the Mackey obstructions of Morita equivalent systems are the same by Proposition 2.1.5, we conclude easily that the Mackey obstruction map is invariant under passing to Morita equivalent twisted actions.

THEOREM 5.5.2. *Suppose that (A, G, α, τ) is a separable abelian twisted system such that A has continuous trace and such that the stabilizer map*

$S : \Omega = \widehat{A} \to \mathfrak{K}(G); x \mapsto S_x$ is continuous. Then $A \rtimes_{\alpha,\tau} G$ has continuous trace if and only if the following conditions are satisfied.

(1) The Mackey obstruction map $h^{\alpha,\tau}$ of (A, G, α, τ) has closed range and is open as a map onto its image.

(2) The canonical action of G on $(A \rtimes_{\alpha,\tau} \Omega^S)\widehat{\ }$ is σ-proper.

We will later see that in some important special cases Condition (1) has an easy description in terms of the Mackey obstructions and the symmetry groups of the system.

Before we start with the proof of the theorem, we need some lemmas. We start with a closer investigation of the Mackey obstruction map $h^{\alpha,\tau}$. For this we first need an alternative description of the topology on $C^*(\Omega^L)\widehat{\ }$ in case where $L : \Omega \to \mathfrak{K}(G)$ is a continuous map and G is abelian.

LEMMA 5.5.3. Let G be an abelian group, and let $L : \Omega \to \mathfrak{K}(G)$ be a continuous map. Then a net $(x_i, \chi_i)_{i \in I} \subseteq C^*(\Omega^L)\widehat{\ }$ converges to $(x, \chi) \in C^*(\Omega^L)\widehat{\ }$ if and only if the following conditions are satisfied:

(1) $x_i \to x$ in Ω.

(2) If $s \in L_x$ and $s_j \in L_{x_{ij}}$ for some subnet $(x_{ij})_{j \in J}$ such that $s_j \to s$ in G, then $\chi_{ij}(s_j) \to \chi(s)$.

PROOF. Observe first that the map

$$\Phi : C_c(\Omega) \odot C_c(\mathfrak{K}(G)^{\mathrm{Id}}) \to C_c(\Omega^L); \Phi(f \otimes g)(x, s) = f(x)g(L_x, s)$$

extends to a $*$-homomorphism from $C_0(\Omega) \otimes C^*(\mathfrak{K}(G)^{\mathrm{Id}})$ onto $C^*(\Omega^L)$ with

$$\ker \Phi = \bigcap \{\ker(x, (L_x, \chi)); x \in \Omega, \chi \in \widehat{L}_x\}.$$

Thus we see that $(x_i, \chi_i) \to (x, \chi)$ if and only if $x_i \to x$ in Ω and $(L_{x_i}, s_i) \to (L_x, s)$ in $\mathfrak{K}(G)^{\mathrm{Id}}$. But by [**24**, Theorem 3.1'] the latter is true if and only if $(\chi_i)_{i \in I}$ satisfies Condition (2). $\square$

We can now use Williams's proposition (Proposition 5.4.4) in order to show that $h^{\alpha,\tau}$ is always continuous.

LEMMA 5.5.4. Let (A, G, α, τ) be a twisted abelian system such that A has continuous trace and such that the stabilizer map $S : \Omega = \widehat{A} \to \mathfrak{K}(G)$ is continuous. Then the Mackey obstruction map $h^{\alpha,\tau} : \Omega^{\tilde{S}} \to \Omega \times_{S^\perp} \widehat{\tilde{G}}$ is continuous.

PROOF. By passing over to a Morita equivalent action we may assume that τ is trivial. Let $(x_n, s_n)_{n \in \mathbb{N}} \subseteq \Omega^S$ be such that $(x_n, s_n) \to (x_0, s_0) \in \Omega^S$. Then $x_n \to x_0$ and it remains to show that Condition (2) of Lemma 5.5.3 is satisfied. So let $t_0 \in S_{x_0}$ and assume that $t_m \in L_{x_{n_m}}$ is such that $t_m \to t_0$ in G for some subsequence $(x_{n_m})_{n \in \mathbb{N}}$ of $(x_n)_{n \in \mathbb{N}}$. By Proposition 5.4.4 we may find representatives ω_y for the Mackey obstructions of y in a neighborhood of $x_0 \in \Omega$ such that $\Omega^S \to \mathbb{T}; (y, s, t) \mapsto \omega_y(s, t)$ is continuous near both, (x_0, s_0, t_0) and (x_0, t_0, s_0). Thus $(x, s, t) \to h^\alpha(x, s)(t) = \omega_x(s, t)\omega_x(t, s)^{-1}$ is also continuous near (x_0, s_0, t_0) from which it immediately follows that $h^\alpha(x_{n_m}, s_{n_m})(t_m) \rightarrow h^\alpha(x_0, s_0)(t_0)$. □

The next lemma gives necessary and sufficient conditions for $h^{\alpha,\tau}$ to have closed range. We need

DEFINITION 5.5.5. If G is a locally compact group and $H, L : \Omega \rightarrow \mathfrak{K}(G)$ are two continuous maps such that $H_x \subseteq L_x$ for all $x \in \Omega$. Then we define

$$\Omega \times_H L = \{(x, \dot{s}) \in \Omega \times_H G; s \in L_x\}.$$

$\Omega \times_H L$ is called the *quotient of Ω^L by Ω^H*.

Note that $\Omega \times_H L$ is the image of Ω^L under the previously defined equivalence relation $\sim_H$ on $\Omega \times G$. Since Ω^L is a closed union of $\sim_H$-equivalence classes, it follows that $\Omega \times_H L$ is closed in $\Omega \times_H G$.

LEMMA 5.5.6. *Let (A, G, α, τ) be a separable twisted abelian system such that A has continuous trace and the stabilizer map $S : \Omega = \widehat{A} \rightarrow \mathfrak{K}(G)$ is continuous. Then $h^{\alpha,\tau} : \Omega^{\tilde{S}} \rightarrow \Omega \times_{S^\perp} \widehat{G}$ has closed range if and only if all Mackey obstructions of (A, G, α, τ) are type I and the symmetrizer map $\Sigma : \Omega \rightarrow \mathfrak{K}(G)$ is continuous. If these conditions are satisfied, then $h^{\alpha,\tau}$ factors through a continuous bijection from $\Omega \times_\Sigma S$ onto $\Omega \times_{S^\perp} \Sigma^\perp$.*

PROOF. As usual we assume that τ is trivial. Since h^α restricted to a fiber S_x is equal to $h_{\omega_x} : S_x \rightarrow \widehat{S}_x$, and since h_{ω_x} has closed range if and only if ω_x is type I by [3], we see that in order for h^α to have closed range, all Mackey obstructions must be type I .

Assume now that all Mackey obstructions are type I but Σ is not continuous. Then, using Corollary 5.4.6, we may find a sequence $(x_n)_{n \in \mathbb{N}} \subseteq \Omega$ converging to $x \in \Omega$ such that Σ_{x_n} converges to a group Σ_0 which is properly contained in Σ_x. Let $\chi' \in \widehat{\Sigma_x/\Sigma_0}$ be any non-trivial character and let χ be an extension of χ' to S_x. Since h_{ω_x} maps S_x into $\widehat{S_x/\Sigma_x}$, we see that χ is not contained in the image of h^α. But χ is contained in the closure of $h^\alpha(\Omega^S)$. In order to see this observe that χ is weakly contained in the left regular representation λ_{S_x/Σ_0} of S_x/Σ_0, and that by the continuity of inducing representations we have

$$(x_n, \lambda_{S_{x_n}/\Sigma_{x_n}}) = (x_n, \text{ind}_{\Sigma_{x_n}}^{S_{x_n}} 1_{\Sigma_{x_n}}) \rightarrow (x, \text{ind}_{\Sigma_0}^{S_x} 1_{\Sigma_0}) = (x, \lambda_{S_x/\Sigma_0})$$

in $\mathcal{S}(\Omega^S)$. But Propositions 1.2.1 and 1.2.3 now imply that, by passing to a subsequence, there exist elements $\chi_n \in \widehat{S_{x_n}/\Sigma_{x_n}}$ such that $(x_n, \chi_n) \rightarrow (x, \chi)$ in $C^*(\Omega^S)\widehat{}$. Since (x_n, χ_n) is in the image of h^α we conclude that χ is contained in the closure of $h^\alpha(\Omega^S)$.

Finally assume that all Mackey obstructions are type I and that $\Sigma : \Omega \rightarrow \mathfrak{K}(G)$ is continuous. Then the image of h^α is exactly $\Omega \times_{S^\perp} \Sigma^\perp$, which is closed in $\Omega \times_{S^\perp} \widehat{G}$. Since Σ_x is the group kernel of h_{ω_x} for each $x \in \Omega$, it follows that h^α factors through a continuous bijection from $\Omega \times_\Sigma S$ onto $\Omega \times_{S^\perp} \Sigma^\perp$. □

It is a consequence of the previous lemma that $h^{\alpha,\tau}$ has closed range and is open as a map onto its image if and only if $h^{\alpha,\tau}$ factors through a homeomorphism between $\Omega \times_\Sigma S$ and $\Omega \times_{S^\perp} \Sigma^\perp$. We see later that in several special situations Conditions (1) and (2) of the lemma are also sufficient for $h^{\alpha,\tau}$ to be open as a map onto its image. In the case $A = \mathcal{K}$ this follows easily from the open mapping theorem for surjective group homomorphisms between σ-compact locally compact groups. But first we proceed with

LEMMA 5.5.7. *Let (A, G, α, τ) be a separable abelian system such that A has continuous trace with $\widehat{A} = \Omega$. Assume that both, the stabilizer map $S : \Omega \to \mathcal{K}(G)$, and the symmetrizer map $\Sigma : \Omega \to \mathcal{K}(G)$ are continuous. Then the map*

$$\mathrm{Ind}_\Sigma^S : (A \rtimes_{\alpha,\tau} \Omega^\Sigma)\widehat{\,} \to \mathrm{Prim}(A \rtimes_{\alpha,\tau} \Omega^S); (x, \rho \times V) \mapsto \ker(x, \mathrm{ind}_{\Sigma_x}^{S_x}(\rho \times V))$$

is a G- and $\widehat{\widehat{G}}$-equivariant homeomorphism with inverse map

$$\mathrm{Res}_\Sigma^S : \mathrm{Prim}(A \rtimes_{\alpha,\tau} \Omega^S) \to \mathrm{Prim}(A \rtimes_{\alpha,\tau} \Omega^\Sigma) \cong (A \rtimes_{\alpha,\tau} \Omega^\Sigma)\widehat{\,}$$

which maps $\ker(x, \pi \times U) \in \mathrm{Prim}(A \rtimes_{\alpha,\tau} \Omega^S)$ to $\ker(x, \pi \times U|_{\Sigma_x}) \in \mathrm{Prim}(A \rtimes_{\alpha,\tau} \Omega^\Sigma)$.

PROOF. Assume that τ is trivial. Since $\mathrm{Prim}(A \rtimes_\alpha \Omega^S) = \bigcup_{x \in \Omega} \mathrm{Prim}(A_x \rtimes_\alpha S_x)$ and $(A \rtimes_\alpha \Omega^\Sigma)\widehat{\,} = \bigcup_{x \in \Omega}(A_x \rtimes_\alpha \Sigma_x)\widehat{\,}$ it follows directly from Proposition 3.1.13 that Ind_Σ^S is a bijection between $(A \rtimes_\alpha \Omega^\Sigma)\widehat{\,}$ and $\mathrm{Prim}(A \rtimes_\alpha \Omega^S)$ with inverse Res_Σ^S. It is trivial to see that Res_Σ^S is equivariant with respect to the canonical actions of G and $\widehat{G}$. Thus the same is true for Ind_Σ^S. Finally, both, Ind_Σ^S and Res_Σ^S are continuous by Proposition 4.1.3. The case of non-trivial τ follows as usual by passing over to a Morita equivalent action. □

We need one final lemma before starting the proof of Theorem 5.5.2.

LEMMA 5.5.8. *Let (A, G, α) be a separable twisted abelian system such that A has continuous trace. Suppose that the stabilizer and symmetrizer maps $S, \Sigma : \Omega = \widehat{A} \to \mathcal{K}(G)$ are continuous, and let γ^Σ denote the canonical action of G on $A \rtimes_\alpha \Omega^\Sigma$. Furthermore, let $(x, \rho \times V) \in (A \rtimes_\alpha \Omega^\Sigma)\widehat{\,}$. Then the stabilizer $S_{(x,\rho \times V)}$ of $(x, \rho \times V)$ is equal to S_ρ and the Mackey obstruction $[\omega_{(x,\rho \times V)}]$ is equal to $[\omega_\rho]$. Moreover, the Mackey obstruction map h^{γ^Σ} for the system $(A \rtimes_\alpha \Omega^\Sigma, G, \gamma^\Sigma)$ has closed range and is open as a map onto its image if and only if the same is true for h^α.*

PROOF. Except of the last assertion about openness, everything follows easily from Lemma 3.3.7. So let $\Lambda = (A \rtimes_\alpha \Omega^\Sigma)\widehat{\,}$, and let $p : \Lambda \to \Omega$ denote the projection given by $(x, \rho \times V) \to x$. The stabilizer and symmetrizer maps for the system $(A \rtimes_\alpha \Omega^\Sigma, G, \gamma^\Sigma)$ are just $S \circ p$ and $\Sigma \circ p$. It follows directly from Lemma

5.5.6 that h^{γ^Σ} has closed range if and only if h^α has closed range. If we define maps $p' : \Lambda \times_{(\Sigma \circ p)} (S \circ p) \to \Omega \times_\Sigma S$ and $p'' : \Lambda \times_{(S \circ p)^\perp} (\Sigma \circ p)^\perp \to \Omega \times_{S^\perp} \Sigma^\perp$ by

$$p'(z, s) = (p(z), s) \quad \text{and} \quad p''(z, \chi) = (p(z), \chi),$$

then p' and p'' are open surjective maps such that the following diagram commutes:

$$
\begin{array}{ccc}
\Lambda \times_{(\Sigma \circ p)} (S \circ p) & \xrightarrow{h^{\gamma^\Sigma}} & \Lambda \times_{(S \circ p)^\perp} (\Sigma \circ p)^\perp \\
{\scriptstyle p'} \downarrow & & \downarrow {\scriptstyle p''} \\
\Omega \times_\Sigma S & \xrightarrow{h^\alpha} & \Omega \times_{S^\perp} \Sigma^\perp.
\end{array}
$$

It follows directly from this diagram that the openness of h^{γ^Σ} implies the openness of h^α. For the converse suppose that h^α is open and let $h^{\gamma^\Sigma}(z_n, s_n) = (z_n, \chi_n) \to (z, \chi) = h^{\gamma^\Sigma}(z, s)$ in $\Lambda \times_{(S \circ p)^\perp} (\Sigma \circ p)^\perp$. Then $(p(z_n), s_n) \to (p(z), s)$ in $\Omega \times_\Sigma S$ by the openness of h^α, from which it follows that $(z_n, s_n) \to (z, s)$ in $\Lambda \times_{\Sigma \circ p} S \circ p$. Thus h^{γ^Σ} is open. $\square$

PROOF OF THEOREM 5.5.2. By passing over to a Morita equivalent action we reduce to the case where τ is trivial. Assume first that $A \rtimes_\alpha G$ has continuous trace. Then it follows from Theorem 3.2.1 that all Mackey obstructions of (A, G, α) must be type I.

By Theorem 3.3.3 we know that the symmetrizer map, say $\tilde{\Sigma}$, for the dual system $(A \rtimes_\alpha G, \widehat{G}, \widehat{\alpha})$ is given by $\tilde{\Sigma}_{\pi \times U} = S^\perp_{\text{Res}^G(\pi \times U)}$, where $\text{Res}^G : (A \rtimes_\alpha G)\widehat{} \to \mathcal{Q}_G(\widehat{A})$ denotes the canonical map. Since S is continuous, it follows that $\tilde{\Sigma}$ is continuous, too. Thus it follows from Theorem 5.2.9 that $(A \rtimes_\alpha G) \rtimes_{\widehat{\alpha}} \widehat{\Omega}^{\tilde{\Sigma}}$ has continuous trace and the action $\widehat{\widehat{\alpha}}$ of G on $((A \rtimes_\alpha G) \rtimes_{\widehat{\alpha}} \widehat{\Omega}^{\tilde{\Sigma}})\widehat{}$ is σ-proper, where $\widehat{\Omega} = (A \rtimes_\alpha G)\widehat{}$.

Let us define $\tilde{\Omega}$ as the closure of the image of S in $\mathfrak{K}(G)$, let $R : \widehat{A} \to \tilde{\Omega}$ be given by $R(\rho) = S_\rho$, and let $L : \tilde{\Omega} \to \mathfrak{K}(G)$ be given by inclusion. Then we have $L \circ R = S \circ \text{id}$, where id denotes the identity on $\widehat{A}$. Moreover, let $(\widehat{R}, L^\perp)$ be the regular pair for $(A \rtimes_\alpha G, \widehat{G}, \widehat{\alpha})$ which is dual to (R, L) (see Definition 4.4.1). Then it is easily seen that $L^\perp \circ \widehat{R} = \tilde{\Sigma} \circ \text{id}$, where here id denotes the identity on $(A \rtimes_\alpha G)\widehat{}$. Thus, combining Theorem 4.4.2 with Proposition 4.1.7, we conclude that the action $\widehat{\widehat{\alpha}}$ of G on $(A \rtimes_\alpha G) \rtimes_{\widehat{\alpha}} \widehat{\Omega}^{\tilde{\Sigma}}$ is Morita equivalent to the canonical action γ^S of G on $A \rtimes_\alpha \Omega^S$. This implies that $A \rtimes_\alpha \Omega^S$ has continuous trace and that G acts σ-properly on $(A \rtimes_\alpha \Omega^S)\widehat{}$.

Now assume that the symmetrizer map $\Sigma : \widehat{A} \to \mathfrak{K}(G)$ is not continuous. Then we can find a sequence $\rho_n \to \rho$ in $\widehat{A}$ such that Σ_{ρ_n} converges to a group, say Σ_0, in $\mathfrak{K}(G)$ with $\Sigma_0 \neq \Sigma_\rho$. We know from Proposition 5.4.6 that $\Sigma_0 \subseteq \Sigma_\rho$. After passing to a subsequence if necessary, we choose $\rho \times V \in (A \rtimes_\alpha \Sigma_\rho)\widehat{}$ and $\rho_n \times V^n \in (A \rtimes_\alpha \Sigma_{\rho_n})\widehat{}$ such that $(\Sigma_{\rho_n}, \rho_n \times V^n) \to (\Sigma_0, \rho \times V|_{\Sigma_0})$ in $C^*(\mathfrak{K}(G)^{\text{Id}}, A, \alpha)\widehat{}$. The existence of such sequence follows from Propositions 1.2.1 and 1.2.3, and the fact that $\text{ind}^L_{\{e\}} \tilde{\rho}$ is weakly equivalent to $(A_{\tilde{\rho}} \rtimes_\alpha L)\widehat{}$ for all $\tilde{\rho} \in \widehat{A}$ and closed subgroups $L \subseteq \Sigma_{\tilde{\rho}}$, together with the fact that by continuity of induction

$$(\Sigma_{\rho_n}, \mathrm{ind}_{\{e\}}^{\Sigma_n} \rho_n) \to (\Sigma_0, \mathrm{ind}_{\{e\}}^{\Sigma_0} \rho)$$

in $\mathcal{S}(\mathfrak{K}(G)^{\mathrm{Id}}, A, \alpha)$. By Proposition 4.1.3 it follows that

$$(S_{\rho_n}, \mathrm{ind}_{\Sigma_{\rho_n}}^{S_{\rho_n}} (\rho_n \times V^n)) \to (S_\rho, \mathrm{ind}_{\Sigma_0}^{S_\rho} (\rho \times V|_{\Sigma_0}))$$

in $\mathcal{S}(\mathfrak{K}(G)^{\mathrm{Id}}, G, \alpha)$, which by an easy application of Proposition 4.1.7 implies that

$$(x_n, \mathrm{ind}_{\Sigma_{\rho_n}}^{S_{\rho_n}} (\rho_n \times V^n)) \to (x, \mathrm{ind}_{\Sigma_0}^{S_\rho} (\rho \times V|_{\Sigma_0}))$$

in $\mathrm{Rep}(A \rtimes_\alpha \Omega^S)$ (here $x_n \in \Omega$ corresponds to ρ_n in $\widehat{A}$). It follows then from Proposition 3.1.13 that for all $n \in \mathbb{N}$, $\ker(x_n, \mathrm{ind}_{\Sigma_{\rho_n}}^{S_{\rho_n}} (\rho_n \times V^n)) \in \mathrm{Prim}(A \rtimes_\alpha \Omega^S)$, while

$$\ker(x, \mathrm{ind}_{\Sigma_0}^{S_\rho} (\rho \times V|_{\Sigma_0})) = \bigcap_{\chi \in \widehat{\Sigma_\rho/\Sigma_0}} \ker(x, \mathrm{ind}_{\Sigma_\rho}^{S_\rho} (\rho \times \chi V)),$$

which follows from

$$\mathrm{ind}_{\Sigma_0}^{S_\rho} (\rho \times V|_{\Sigma_0}) = \mathrm{ind}_{\Sigma_\rho}^{S_\rho} (\mathrm{ind}_{\Sigma_0}^{\Sigma_\rho} (\rho \times V|_{\Sigma_0})) = \mathrm{ind}_{\Sigma_\rho}^{S_\rho} (\lambda_{\Sigma_\rho/\Sigma_0} \otimes (\rho \times V))$$

which is weakly equivalent to $\{\mathrm{ind}_{\Sigma_\rho}^{S_\rho} (\rho \times \chi V); \chi \in \widehat{\Sigma_\rho/\Sigma_0}\}$. But this contradicts the fact that $\mathrm{Prim}(A \rtimes_\alpha \Omega^S)$ is Hausdorff, which is the case since we already know that $A \rtimes_\alpha \Omega^S$ has continuous trace.

In order to show that h^α is an open map onto its image it is, by Lemma 5.5.8, enough to show that this is true for h^{γ^Σ}, where γ^Σ denotes the canonical action of G on $A \rtimes_\alpha \Omega^\Sigma$. For this let $\Lambda = (A \rtimes_\alpha \Omega^\Sigma)^{\widehat{\ }}$, and observe that for each $z \in \Lambda$, h^{γ^Σ} is given on the fiber S_z/Σ_z by the isomorphism $h_{\omega_z} : S_z/\Sigma_z \to \widehat{\Sigma_z^\perp/S_z^\perp}$, whose inverse is given by $h_{\widehat{\omega_z}} : \widehat{\Sigma_z^\perp/S_z^\perp} \to S_z/\Sigma_z$, where $\widehat{\omega}_z$ is the dual multiplier of ω_z as defined in Definition 3.3.4. Thus it is a consequence of Theorem 3.3.6 and Lemma 5.5.8 that the map $h^{\gamma^{S^\perp}}$ belonging to the system $((A \rtimes_\alpha G) \rtimes_{\widehat{\alpha}} \widehat{\Omega}^{S^\perp}, \widehat{G}, \gamma^{S^\perp})$ is inverse to h^{γ^Σ}. Since $A \rtimes_\alpha G$ has continuous trace by assumption, it follows now from Lemma 5.5.4 that $h^{\gamma^{S^\perp}}$ is continuous, which just means that h^{γ^Σ} is open.

Assume now that (A, G, α) satisfies Conditions (1) and (2) of the theorem. We reduce to the case where G acts σ-properly on $\widehat{A}$. For this observe that by Condition (2) and Lemma 5.5.7, G acts σ-properly on $(A \rtimes_\alpha \Omega^\Sigma)^{\widehat{\ }}$. Since $A \rtimes_\alpha G$ is a quotient of $(A \rtimes_\alpha \Omega^\Sigma) \rtimes_{\gamma^\Sigma} G$ by Proposition 4.3.5, it is enough to show that the latter crossed product has continuous trace in order to show that $A \rtimes_\alpha G$ has continuous trace. Moreover, by Lemma 5.5.8 we know that Condition (1) is satisfied for the iterated system $(A \rtimes_\alpha \Omega^\Sigma, G, \gamma^\Sigma)$. Thus we may assume without loss of generality that (A, G, α) itself is an abelian system which satisfies Condition (1) such that G acts σ-properly on $\widehat{A}$.

In a next step we are going to show that (A, G, α) has continuously varying maximally pointwise unitary subgroups. For this let $(x_n)_{n\in\mathbb{N}}$ be a sequence in Ω which converges to $x \in \Omega$, and let H_n be a maximally x_n-unitary subgroup of G. By Corollary 5.4.5 we may assume that $H_n \to H_0$ in $\mathfrak{K}(G)$ such that H_0 is contained in a maximally x-unitary subgroup, say H, of G. We show that $H_0 = H$. For this recall that by Condition (1) h^α may be viewed as a homeomorphism $h^\alpha : \Omega \times_\Sigma S \to \Omega \times_{S^\perp} \Sigma^\perp$ with inverse given by the map, say $h^{\tilde\alpha}$, which is given on the fibers by $h_{\widehat{\omega_x}}$. In particular this implies that $h^{\tilde\alpha}$, viewed as function from $\Omega^{\Sigma^\perp}$ into $\Omega \times_\Sigma G$, is continuous. Let $\chi, \mu \in H_0^\perp$. By passing to a subsequence if necessary, we find elements $\chi_n, \mu_n \in H_n^\perp$ such that $\chi_n \to \chi$ and $\mu_n \to \mu$ in $\widehat{G}$. Using Lemma 5.5.3, the continuity of $h^{\tilde\alpha}$ implies that

$$1 = h^{\tilde\alpha}(x_n, \chi_n)(\mu_n) \to h^{\tilde\alpha}(x, \chi)(\mu),$$

since for each $n \in \mathbb{N}$, $H_n^\perp$ is a maximally $\widehat{\omega}_{x_n}$-trivial subgroup of $\widehat{G}$. It follows that the restriction of $\widehat{\omega}_x$ to $H_0^\perp$ is trivial. But if $H_0 \neq H$ (i.e. $H^\perp \subsetneqq H_0^\perp$), this contradicts the fact that $H^\perp$ is a maximally $\widehat{\omega}_x$-trivial subgroup of $\widehat{G}$.

Since G acts σ-properly on $\widehat{A}$ we conclude that $\widehat{A}/G$ is Hausdorff. Thus by Theorem 5.4.3 it remains to show that for any continuous subchoice of maximally pointwise unitary subgroups $H : M \to \mathfrak{K}(G)$ for (A, G, α) the action of G on $(A_M \rtimes_\alpha M^H)^\widehat{\ }$ is σ-proper. For this we may of course assume that $M = \Omega$. Let $\Lambda = (A \rtimes_\alpha \Omega^H)^\widehat{\ }$ and let $((x_n, \rho_n \times V^n), s_n)_{n \in \mathbb{N}}$ be a sequence in $\Lambda \times G$ such that $((x_n, \rho_n \times V^n), s_n \cdot (x_n, \rho_n \times V^n))$ converges to some pair $((x, \rho \times V), (x', \rho' \times V')) \in \Lambda \times \Lambda$. Then we have to show, by passing to a subsequence if necessary, that there exists a sequence $h_n \in H_{x_n}$ such that $((x_n, \rho_n \times V^n), s_n h_n)_{n \in \mathbb{N}}$ converges in $\Lambda \times G$. Since $(x_n, \rho_n \times V^n) \to (x, \rho \times V)$ in Λ, this clearly amounts to showing that $(s_n h_n)_{n \in \mathbb{N}}$ is a convergent sequence in G. For this let us first observe that $x_n \to x$ and $s_n x_n \to x'$ in Ω. By the σ-properness of the G-action on Ω we may therefore assume that there exist elements $t_n \in S_{x_n}$ such that $(x_n, s_n t_n) \to (x, s)$ in $\Omega \times G$ for some $s \in G$. By multiplying our original sequence $((x_n, \rho_n \times V^n), s_n)_{n \in \mathbb{N}}$ with the convergent sequence $(t_n^{-1} s_n^{-1})_{n \in \mathbb{N}}$ in both components, we reduce everything to the case where each s_n is an element of S_{x_n}.

Recall now that $s_n \cdot (x_n, \rho_n \times V^n) = (x_n, \rho_n \times h_{\omega_{x_n}}(s_n^{-1}) V^n)$ for all $n \in \mathbb{N}$ by Lemma 3.1.10. Extending each $h_{\omega_{x_n}}(s_n^{-1})$ to a character χ_n of G, we obtain a sequence $((x_n, \rho_n \times V^n), \chi_n)_{n \in \mathbb{N}} \subseteq \Lambda \times \widehat{G}$ such that $((x_n, \rho_n \times V^n), \chi_n(x_n, \rho_n \times V^n))_{n \in \mathbb{N}}$ converges in $\Lambda \times \Lambda$. Since by Proposition 5.2.9 the action of $\widehat{G}$ on Λ is σ-proper, we may alter each χ_n by multiplying an element, say $\mu_n \in H_{x_n}^\perp$ such that $(\chi_n \mu_n)_{n \in \mathbb{N}}$ converges to some χ in $\widehat{G}$. This implies that $(x_n, \chi_n \mu_n|_{S_{x_n}})$ converges to $(x, \chi|_{S_x})$ in $\Omega \times_{S^\perp} \widehat{G}$. Since the image of H_{x_n} under $h_{\omega_{x_n}}$ equals $\widehat{S_{x_n}/H_{x_n}}$ by the type I'ness of the ω_{x_n}'s, we find $h_n \in H_{x_n}$ such that $\mu_n|_{S_{x_n}} = h_{\omega_{x_n}}(h_n^{-1})$. Thus we conclude that $(x_n, h_{\omega_{x_n}}(h_n^{-1} s_n^{-1})) = h^\alpha(x_n, h_n^{-1} s_n^{-1})$ is a convergent sequence in $\Omega \times_{S^\perp} \widehat{G}$. Now the openness of h^α implies that we may alter each h_n by multiplication with an element in $\Sigma_{x_n} \subseteq H_{x_n}$ (after passing to a subsequence if necessary) such that $(x_n, h_n^{-1} s_n^{-1})_{n \in \mathbb{N}}$ converges in $\Omega^S \subseteq \Omega \times G$. But then $(h_n s_n)_{n \in \mathbb{N}}$ is a convergent sequence in G, which finishes the proof. $\square$

We proceed with the following corollary of Theorem 5.5.2.

COROLLARY 5.5.9. *Assume that (A, G, α, τ) is a separable twisted abelian system such that G acts σ-properly on $\Omega = \widehat{A}$ (in particular, this is true if G acts trivially on $\widehat{A}$). Then $A \rtimes_{\alpha,\tau} G$ has continuous trace if and only if the Mackey obstruction map $h^{\alpha,\tau} : \Omega^{\tilde{S}} \to \Omega \times_{S^{\perp}} \widehat{G}$ has closed range and is open as a map onto its image.*

The proof of the corollary follows directly from Theorem 5.5.2 as soon as we have shown that G acts σ-properly on $(A \rtimes_{\alpha,\tau} \Omega^S)\widehat{}$ if $h^{\alpha,\tau}$ has closed range. But this follows from Lemma 5.5.6 together with Lemma 5.5.7 and

LEMMA 5.5.10. *Suppose that (A, G, α, τ) is separable abelian twisted system such that A has continuous trace and G acts σ-properly on $\Omega = \widehat{A}$. Assume further that the symmetrizer map $\Sigma : \Omega \to \mathfrak{K}(G)$ is continuous. Then the canonical action of G an $(A \rtimes_{\alpha,\tau} \Omega^{\Sigma})\widehat{}$ is σ-proper, too.*

PROOF. By passing over to a Morita equivalent action we may assume that τ is trivial (see Proposition 4.4.6). Let $\Lambda = (A \rtimes_{\alpha} \Omega^{\Sigma})\widehat{}$ and let $((x_n, \rho_n \times V^n), s_n)_{n \in \mathbb{N}} \subseteq \Lambda \times G$ such that $((x_n, \rho_n \times V^n), s_n \cdot (x_n, \rho_n \times V^n))_{n \in \mathbb{N}}$ converges in $\Lambda \times \Lambda$. Then it is enough to show that, after probably passing to a subsequence, there exist elements $t_n \in S_{x_n}$ such that $(t_n s_n)_{n \in \mathbb{N}}$ converges in G. But the convergence of $((x_n, \rho_n \times V^n), s_n \cdot (x_n, \rho_n \times V^n))_{n \in \mathbb{N}}$ in $\Lambda \times \Lambda$ implies the convergence of $(x_n, s_n x_n)_{n \in \mathbb{N}}$ in $\Omega \times \Omega$ from which follows that $(x_n, \dot{s}_n)$ has a convergent subsequence in $\Omega \times_S G$. Thus, the result follows by the openness of the quotient map $\Omega \times G \to \Omega \times_S G$. $\square$

If $\tilde{G} = G/N_{\tau}$ is compact we get the following result.

COROLLARY 5.5.11. *Let (A, G, α, τ) be an abelian separable twisted covariant system such that A has continuous trace and such that $\tilde{G} = G/N_{\tau}$ is compact. Suppose further that the stabilizer map $S : \widehat{A} \to \mathfrak{K}(G)$ is continuous. Then $A \rtimes_{\alpha,\tau} G$ has continuous trace.*

PROOF. By localizing the problem, we may as well assume that $\Omega = \widehat{A}$ is compact. Since $\tilde{G}$ is compact, the continuity of the stabilizer map implies that G acts σ-properly on Ω. Thus the corollary follows from Corollary 5.5.9 if we can show that $h^{\alpha,\tau}$ has closed range and is open as a map onto its image. But this follows directly from the fact that $h^{\alpha,\tau}$ is continuous and that $\Omega^{\tilde{S}} \subseteq \Omega \times \tilde{G}$ is compact. $\square$

Note that even in the case where $\tilde{G}$ is finite, it is not necessary that the stabilizer map is continuous for $A \rtimes_{\alpha,\tau} G$ to have continuous trace (see Example 6.3.2 below). As an interesting consequence of the proof given above, together with Lemma 5.5.6, we see that in case where $\tilde{G}$ is compact, the continuity of the stabilizer map implies the continuity of the symmetrizer map. This is not very surprising if we keep in mind that in the special situation where all stabilizers are equal to G the map $\widehat{A} \to H^2(\tilde{G}, \mathbb{T}); \rho \mapsto [\omega_{\rho}]$ is continuous [**46**, Lemma 3.3], and hence locally constant, since the second Moore cohomology group of a compact group is discrete [**43**]. However, many examples show that a similar result is not true for actions of arbitrary groups. For instance, if G is the Heisenberg group, that is $G = \mathbb{R}^3$ equipped with multiplication given by $(s, t, r)(s', t', r') = (s+s', t+t', r+r'+st')$, then we may write $C^*(G)$ as the twisted crossed product $C^*(Z) \rtimes_{\gamma^Z, \tau^Z} G$ by decomposing with respect to the center $Z = (0, 0, \mathbb{R}) \subseteq G$. All stabilizers of the system are equal to G, but the symmetrizer map of the system is not continuous since $\Sigma_\lambda = Z$ for all $0 \neq \lambda \in \mathbb{R} \cong C^*(Z)\widehat{\ }$ but $\Sigma_0 = G$. However, in general it is certainly interesting to ask under which conditions on (A, G, α, τ) the following conjecture is true, which, if true, would generalize the well known fact that for any multiplier ω on a separable abelian group G, $h_\omega : G \to \widehat{G}$ has closed range and is open as a map onto its image if and only if ω is type I.

CONJECTURE 5.5.12. *Let* (A, G, α, τ) *be a separable abelian twisted system such that* A *has continuous trace and such that the stabilizer map* $S : \Omega = \widehat{A} \to \mathcal{R}(G)$ *is continuous. Then the Mackey obstruction map* $h^{\alpha, \tau} : \Omega^{\tilde{S}} \to \Omega \times_{S^\perp} \widehat{\widehat{G}}$ *has closed range and is open as a map onto its image if and only if all Mackey obstructions of* (A, G, α, τ) *are type I and the symmetrizer map* $\Sigma : \Omega \to \mathcal{R}(G)$ *is continuous.*

Clearly, the only if direction of this conjecture follows from Lemma 5.5.6. We do not know any counterexample for Conjecture 5.5.12. In order to give a little evidence for our conjecture we are going to prove

THEOREM 5.5.13. *Let* (A, G, α, τ) *be a separable twisted abelian system such that* A *has continuous trace. Assume further that* (A, G, α, τ) *has constant stabilizer* S *such that* $\tilde{S} = S/N_\tau$ *is a compactly generated Lie group. Then Conjecture 5.5.12 is true for* (A, G, α, τ).

As a direct consequence of Theorem 5.5.13, Theorem 5.5.2 and Corollary 5.5.9 we obtain

COROLLARY 5.5.14. *Let* (A, G, α, τ) *be as in Theorem 5.5.13. Then* $A \rtimes_{\alpha, \tau} G$ *has continuous trace if and only if the following conditions are satisfied:*
  (1) *All Mackey obstructions of* (A, G, α, τ) *are type I.*
  (2) *The symmetrizer map* $\Sigma : \Omega \to \mathcal{R}(G)$ *is continuous.*
  (3) *The action of* G *on* $(A \rtimes_{\alpha, \tau} S)\widehat{\ }$ *factors through a proper action of* G/S.
If, moreover, $\widehat{A}$ *is a proper* G/S-space, then $A \rtimes_{\alpha, \tau} G$ *has continuous trace if and only if Conditions (1) and (2) are satisfied.*

Remark 5.5.15. If ω is a multiplier on the abelian locally compact group G, and if G is compactly generated or a Lie group, then ω is type I if and only if G/Σ_ω may be written as a direct product $V \times Z \times T \times F$, where V is a vector group, Z is a finitely generated free abelian group, T is a torus group with dimension equal to the rank of Z, and F is finite (see [**19**, Lemma 2]). Thus the answer to the question whether ω is type I or not only depends on the structure of G/Σ_ω.

Another situation where we know that Conjecture 5.5.12 is true is the following

THEOREM 5.5.16. *Suppose that (A, G, α, τ) is a separable abelian twisted system such that A has continuous trace. Assume that $\tilde{G} = G/N_\tau$ and $\tilde{S}_\rho = S_\rho/N_\tau$ are vector groups for all $\rho \in \hat{A}$. Then Conjecture 5.5.12 is true for (A, G, α, τ).*

Before we start with the proof of the theorems above, it is useful to prove some lemmas about convergence of subgroups of abelian groups.

LEMMA 5.5.17. *Suppose that G is a locally compact group. If N is an open normal subgroup of G and $q : G \to G/N$ denotes the quotient map, then the maps $q_* : \mathfrak{K}(G) \to \mathfrak{K}(G/N); H \mapsto q(H)$ and $\cap_N : \mathfrak{K}(G) \to \mathfrak{K}(N); H \mapsto H \cap N$ are continuous. Moreover, if G is abelian and K is a compact subgroup of G, then $\cap_K : \mathfrak{K}(G) \to \mathfrak{K}(K); H \to H \cap K$ is continuous, too.*

PROOF. The continuity of the maps q_* and $\cap_N$ follows almost directly from the definition of the topology on $\mathfrak{K}(G)$. To prove the second assertion note that $\cap_K$ can be obtained as the composition of the maps

$$\mathfrak{K}(G) \to \mathfrak{K}(\hat{G}); H \mapsto H^\perp,$$

$$\mathfrak{K}(\hat{G}) \to \mathfrak{K}(\hat{K}); H^\perp \mapsto q^{\hat{K}}(H^\perp),$$

where $q^{\hat{K}} : \hat{G} \to \hat{K}$ denotes the quotient map, and

$$\mathfrak{K}(\hat{K}) \to \mathfrak{K}(K); L^\perp \mapsto (L^\perp)^\perp \ (= L),$$

which are all continuous. $\square$

The next lemma gives some information about convergent sequences of subgroups of vector groups. Recall that every closed subgroup H of a vector group V can be written as a direct sum $H = W \oplus Z$ of a vector subgroup W of V and a subgroup Z of V which is a finitely generated free abelian group. Thus, for any subgroup H of V there are two characterizing natural numbers (including zero), namely the dimension, $\dim H = \dim W$, and the rank, $\operatorname{rank} H = \operatorname{rank} Z$. Note that if V_0 denotes the linear hull of H in V, then $\dim V_0 = \dim H + \operatorname{rank} H$.

LEMMA 5.5.18. *Let V be a vector group, and let $H_n \to H$ in $\mathfrak{K}(V)$. If $\operatorname{rank} H \geq \operatorname{rank} H_n$ for all $n \in \mathbb{N}$, then there exists a subsequence $(H_{n_k})_{k \in \mathbb{N}}$ of $(H_n)_{n \in \mathbb{N}}$ and elements $c_k \in \mathrm{GL}(V)$ such that $c_k \to 1_V$ and $c_k(H) = H_{n_k}$ for all $k \in \mathbb{N}$.*

PROOF. Let $H = W \oplus Z$ be a decomposition of H as above. Let $\{s_1, \ldots, s_r\}$ be a base of W and let $\{s_{r+1}, \ldots, s_l\}$, $l \geq r$, be a minimal set of generators of Z. Moreover, let $\{s_{l+1}, \ldots, s_m\} \subseteq V \smallsetminus H$ such that $\{s_1, \ldots, s_m\}$ becomes a base of V. For each $k \in \mathbb{N}$ let U_k denote the open ball of radius $\frac{1}{k}$. Then it follows from the definition of the topology on $\mathfrak{K}(V)$ that, for all k, there exist elements $n_k \in \mathbb{N}$ such that $H_{n_k} \cap (s_j + U_k) \neq \emptyset$ for all $j = 1, \ldots, m$. For each k choose $s_j^k \in H_{n_k} \cap (s_j + U_k)$. Let $k_0 \in \mathbb{N}$ such that $\{s_1^k, \ldots, s_m^k\}$ are linearly independent for $k \geq k_0$, and, for those k, define $c_k \in \mathrm{GL}(V)$ by $c_k(s_j) = s_j^k$. It is clear that $c_k \to 1_V$ in $\mathrm{GL}(V)$, and it remains to show that $c_k(H) = H_{n_k}$ for all $k \geq k_0$.

We assume first that all H_n are vector groups. Then H is a vector group, too, which follows easily from Proposition 4.1.1. We claim that $\dim H_n = \dim H$ for all but finitely many $n \in \mathbb{N}$. For this let $\{s_1, \ldots, s_r\}$ be the base of H as chosen above. Then, for k big enough, we have seen that $\{s_1^k, \ldots, s_r^k\}$ is a linear independent subset of H_{n_k}, from which follows that $\dim H_{n_k} \geq \dim H$ for almost all k. Using the same argument for any subsequence of $(H_n)_{n \in \mathbb{N}}$ we see that $\dim H_n \geq \dim H$ for all but finitely many n's. Since $H_n^\perp \to H^\perp$ in $\mathfrak{K}(\widehat{V})$, the same argument implies that $\dim H_n \leq \dim H$ for almost all $n \in \mathbb{N}$, which proves the claim.

Next we show that $\dim H + \mathrm{rank}\, H \leq \dim H_n + \mathrm{rank}\, H_n$ for almost all n. For this let V_n be the linear hull of H_n in V. By the compactness of $\mathfrak{K}(V)$ we may assume that V_n converges to some vector subgroup V_0 of V. It is clear that $H \subseteq V_0$. Thus, $\dim H + \mathrm{rank}\, H \leq \dim V_0 = \dim V_n$ for almost all $n \in \mathbb{N}$. Since we may assume without loss of generality that $\dim V_n = \dim H_n + \mathrm{rank}\, H_n$ is constant for all but finitely many $n \in \mathbb{N}$, (otherwise we repeat the argument to any subsequence with constant sums of dimension and rank) we conclude that $\dim H + \mathrm{rank}\, H \leq \dim H_n + \mathrm{rank}\, H_n$ for almost all $n \in \mathbb{N}$.

Now, for each $k \in \mathbb{N}$, let W_k denote the vector group part of H_{n_k} and let V_k be the linear hull of $\{s_{r+1}^k, \ldots, s_l^k\}$. We claim that $W_k \cap V_k = \{0\}$ for almost all $k \in \mathbb{N}$. For this we observe that V_k converges to the linear hull, say $\tilde{V}$, of $s_{r+1}, \ldots, s_l$. If there is a subsequence $(H_{n_j})_{j \in \mathbb{N}}$ of $(H_{n_k})_{k \in \mathbb{N}}$ such that $W_{n_j} \cap V_{n_j} \neq \{0\}$, then we may assume that a subsequence of this subsequence converges to a nontrivial vector subgroup of $H \cap \tilde{V} = Z$. But this is impossible since Z is discrete. It follows from this that $\{s_{r+1}^k, \ldots, s_l^k\}$ generates a discrete subgroup of H_{n_k} which has trivial intersection with W_k.

We conclude from this that $\mathrm{rank}\, H_{n_k} \geq \mathrm{rank}\, H$ for all but finitely many k's, which implies that $\mathrm{rank}\, H_{n_k} = \mathrm{rank}\, H$ for all those k's by assumption. But this implies that $\{s_1^k, \ldots, s_r^k\} \subseteq W_k$ for almost all k, since otherwise we would have $\mathrm{rank}\, H_{n_k} > \mathrm{rank}\, H$. It follows that $\dim W_k \geq \dim W$ for almost all k. On the other side we may assume that $\dim W \geq \dim W_k$, since at least a subsequence of $(W_k)_{k \in \mathbb{N}}$ converges to a vector subgroup of W with dimension equal to $\dim W_k$ (by passing to a subsequence we may of course assume that $\dim W_k$ is constant). Thus we see that, after passing to a subsequence if necessary, $W_k = c_k(W)$ for all $k \in \mathbb{N}$.

In order to see that $c_k(H) = H_{n_k}$ for all k it is now enough to show that $c_k^{-1}(H_{n_k})/W = H/W$. For this observe that $(c_k^{-1}(H_{n_k})/W)_{k \in \mathbb{N}}$ is a sequence of discrete subgroups of V/W such that $c_k^{-1}(H_{n_k})/W \to H/W$ and such that $H/W \subseteq c_k^{-1}(H_{n_k})/W$ for all k. Since the linear hull of all $c_k^{-1}(H_{n_k})$ is constant for all k, which follows from the fact that all these groups have constant rank, we may assume that V/W is equal to this linear hull. Thus we conclude that $(c_k^{-1}(H_{n_k})/W)^\perp \subseteq (H/W)^\perp$ forms a sequence of discrete groups in $\widehat{V/W}$ which converges to $(H/W)^\perp$. But this implies that all these groups do coincide. $\square$

PROOF OF THEOREM 5.5.13. By Lemma 5.5.6 it remains to show that the continuity of the symmetrizer map and the type I'ness of the Mackey obstructions implies that $h^{\alpha,\tau}$ is open as a map onto its image. By passing over to a Morita equivalent action we may assume that τ is trivial. Moreover, by the definition of h^α we may also assume that the constant stabilizer S is equal to G. Each compactly generated Lie group is a quotient of a group $V \times Z$ by some discrete subgroup D, where V is a vector group and Z is a finitely generated free abelian group. Thus, by extending the Mackey obstructions to this covering group, we assume without loss of generality that $G = V \times Z$.

Let $(x_n, s_n)_{n \in \mathbb{N}}$ be a sequence in $\Omega \times G$ and $(x, s) \in \Omega \times G$ such that $h^\alpha(x_n, s_n) \to h^\alpha(x, s)$ in $\Omega \times \widehat{G}$. We have to show that, after passing to a subsequence if necessary, there exist elements $t_n \in \Sigma_{x_n}$ such that $(x_n, t_n s_n) \to (x, s)$ in $\Omega \times G$. It is clear that $x_n \to x$ in Ω. Let $q : G \to G/V$ denote the quotient map. By Lemma 5.5.17 we know that $q(\Sigma_{x_n}) \to q(\Sigma_x)$ and $\Sigma_{x_n} \cap V \to \Sigma_x \cap V$. Let $\text{rank}(\Sigma_{x_n} \cap V)$ be defined as in the remarks preceding Lemma 5.5.18. We claim that $\text{rank}(\Sigma_{x_n} \cap V) \leq \text{rank}(\Sigma_x \cap V)$. In order to see this observe that $\widehat{G} = \widehat{V} \times \widehat{Z}$, where $\widehat{Z}$ is a torus group. By Lemma 5.5.17 we know that $\Sigma_{x_n}^\perp \cap \widehat{Z} \to \Sigma_x^\perp \cap \widehat{Z}$, from which follows that $\Sigma_{x_n}^\perp \cap \widehat{Z} \subseteq \Sigma_x^\perp \cap \widehat{Z}$ for almost all $n \in \mathbb{N}$ (since $\widehat{Z}$ is a torus group). Thus the Lie group dimension of $\Sigma_{x_n}^\perp \cap \widehat{Z}$ is less than or equal to the Lie group dimension of $\Sigma_x^\perp \cap \widehat{Z}$. But it is easily seen that the Lie group dimension of $H^\perp \cap \widehat{Z}$ is equal to $\text{rank}(H \cap V)$ for any closed subgroup H of G which satisfies that G/H is isomorphic to $H^\perp$ (since $\text{rank}(H \cap V)$ is equal to the dimension of the torus in the splitting of G/H). Hence, it follows that

$$\text{rank}(\Sigma_{x_n} \cap V) = \dim(\Sigma_{x_n}^\perp \cap \widehat{Z}) \leq \dim(\Sigma_x^\perp \cap \widehat{Z}) = \text{rank}(\Sigma_x \cap V)$$

for almost all $n \in \mathbb{N}$, which proves the claim.

Applying Lemma 5.5.18 to the sequence $\Sigma_{x_n} \cap V$ we conclude, by passing to a subsequence if necessary, that there exist elements $c_n \in \text{GL}(V)$ such that $c_n \to 1_V$ and $c_n(\Sigma_x \cap V) = \Sigma_{x_n} \cap V$ for all $n \in \mathbb{N}$. In particular we conclude from this that the Lie group dimension of $\Sigma_{x_n}^\perp \cap \widehat{Z}$ is constantly equal to the Lie group dimension of $\Sigma_x^\perp \cap \widehat{Z}$ for all $n \in \mathbb{N}$, from which easily follows that all these groups do coincide. But this implies that

$$q(\Sigma_{x_n}) = (\Sigma_{x_n}^{\perp} \cap \widehat{Z})^{\perp} = (\Sigma_x^{\perp} \cap \widehat{Z})^{\perp} = q(\Sigma_x)$$

for all $n \in \mathbb{N}$, where the annihilators are taken in Z.

For each $n \in \mathbb{N}$ we now define elements $\sigma_n \in \mathrm{Aut}(G)$ by $\sigma_n = c_n \times \mathrm{id}_Z$, where id_Z denotes the identity on Z. Then $\sigma_n \to \mathrm{id}_G$ in $\mathrm{Aut}(G)$ with respect to the compact open topology. Moreover, we have

$$\sigma_n(\Sigma_x \cap V) = \Sigma_{x_n} \cap V \quad \text{and} \quad q(\sigma_n(\Sigma_x)) = q(\Sigma_{x_n})$$

for all $n \in \mathbb{N}$. Now let $M = \{x_n; n \in \mathbb{N}\} \cup \{x\}$. We define a map $\tilde{h} : M \times G \to M \times \widehat{G}$ by

$$\tilde{h}(x_n, s) = (x_n, h_{\omega_{x_n} \circ \sigma_n}(s))$$

for $n \in \mathbb{N}$ and $\tilde{h}(x, s) = h^{\alpha}(x, s)$ for all $s \in G$, where for any multiplier ω on G and $\sigma \in \mathrm{Aut}(G)$, $\omega \circ \sigma(s, t) = \omega(\sigma(s), \sigma(t))$.

It is trivially seen that $\tilde{h}$ is continuous and we proceed by showing that it is open as a map onto its image. For this we put $\tilde{h}_{x_n} = h_{\omega_{x_n} \circ \sigma_n}$ for all $n \in \mathbb{N}$ and $\tilde{h}_x = h_{\omega_x}$. Furthermore, for $n \in \mathbb{N}$ let $\tilde{\Sigma}_{x_n} = \sigma_n^{-1}(\Sigma_{x_n}) = \ker \tilde{h}_{x_n}$ and let $\tilde{\Sigma}_x = \Sigma_x = \ker \tilde{h}_x$. Then $\tilde{\Sigma}_{x_n} \cap V = \Sigma_x \cap V$ for all $n \in \mathbb{N}$, and we may pass over to $G/(\Sigma_x \cap V)$ in order to show the openness of $\tilde{h}$. After having done this, we may assume that $\tilde{\Sigma}_{x_n}$ has trivial intersection with the connected component G_0 of G for all $n \in \mathbb{N}$, and the same may be assumed for $\tilde{\Sigma}_x$. Moreover, if $q : G \to G/G_0$ denotes the quotient map, then it follows from our constructions that $q(\tilde{\Sigma}_{x_n}) = q(\tilde{\Sigma}_x)$ for all $n \in \mathbb{N}$.

Let $H = \tilde{\Sigma}_x \cdot G_0$. Then $H = \tilde{\Sigma}_y G_0$ for all $y \in M$ and we conclude that

$$\tilde{h}_y(sH) = \tilde{h}_y(s)\tilde{h}_y(\tilde{\Sigma}_y G_0) = \tilde{h}_y(s)\tilde{h}_y(G_0)$$

lies in the connected component of $\widehat{G}$ if and only if $s \in H$. On the other hand, since each $\tilde{\Sigma}_y$ has trivial intersection with G_0 for all $y \in M$, we see that each $\tilde{h}_y$ maps G_0 onto the connected component of $\widehat{G}$. Thus, we conclude that the image of $M \times G_0$ under $\tilde{h}$ is open in $M \times \widehat{G}$. Suppose now that $(s_n)_{n \in \mathbb{N}}$ is a sequence in G such that $\tilde{h}(x_n, s_n) \to \tilde{h}(x, s)$ for some $s \in G$. By multiplying this sequence with $s^{-1} z_n$ in the second variable, for appropriate elements $z_n \in \tilde{\Sigma}_{x_n}$, we may assume that $s_n \in G_0$ for all $n \in \mathbb{N}$ and that $\tilde{h}(x_n, s_n) \to (x, 1_G)$ in $\Omega \times \widehat{G}$. We want to show that $s_n \to e$ in G. For this we write $G_0 = W \times T$ for a vector group W and a torus group T, and we write $s_n = r_n \cdot t_n$, $r_n \in W$ and $t_n \in T$. By the compactness of T we may assume that $t_n \to t$ for some $t \in T$. It follows that $s_n t_n^{-1} \in W$ for all $n \in \mathbb{N}$, and that $\tilde{h}_{x_n}(s_n t_n^{-1})|_W \to 1_W$ in $\widehat{W}$. Since each $\tilde{\Sigma}_{x_n}$ has trivial intersection with G_0, and hence also with W, we see that the restriction of $\omega_{x_n} \circ \sigma_n$ to $W \times W$ is totally skew for all $n \in \mathbb{N}$, and the same applies to the restriction of ω_x to $W \times W$. Thus, for each $y \in M$, we obtain an isomorphism $W \to \widehat{W}$ given by $s \mapsto \tilde{h}_y(s)|_W$. By identifying W with $\widehat{W}$ via any fixed linear isomorphism, the continuity of $\tilde{h}$ implies that the map $M \to \mathrm{GL}(W); y \mapsto \tilde{h}_y|_W$ is continuous, too. Thus, the map $M \to \mathrm{GL}(W); y \mapsto (\tilde{h}_y|_W)^{-1}$ is also continuous.

This implies that $s_n t_n^{-1} \to \{0\}$ in W, which in turn yields $s_n \to t$ in G. But since by assumption $\tilde{h}_{x_n}(s_n) \to \tilde{h}_x(t) = 1_G$ in $\widehat{G}$ and since $\ker \tilde{h}_x \cap G_0 = \Sigma_x \cap G_0 = \{e\}$ it follows that $t = 1$. This completes the proof for the openness of $\tilde{h}$.

Finally, let $(s_n)_{n \in \mathbb{N}}$ be a sequence in G such that $h^\alpha(x_n, s_n) \to h^\alpha(x, s)$ for some $s \in G$. Then it follows that $\tilde{h}(x_n, \sigma_n^{-1}(s_n)) = (x_n, h_{\omega_{x_n}}(s_n) \circ \sigma_n)$ converges to $(x, h_{\omega_x}(s)) = \tilde{h}(x, s)$ in $M \times \widehat{G}$. Since $\tilde{h}$ is open as a map onto its image, we may multiply each $\sigma_n(s_n)$ with an appropriate element $r_n \in \tilde{\Sigma}_{x_n} = \sigma_n^{-1}(\Sigma_{x_n})$ such that, after possibly passing to a subsequence, $(x_n, \sigma_n^{-1}(s_n) r_n) \to (x, s)$ in $\Omega \times G$. Thus, if $r_n' = \sigma_n(r_n)$, we have $r_n' \in \Sigma_{x_n}$ for all $n \in \mathbb{N}$ and $(x_n, s_n r_n') \to (x, s)$ in $\Omega \times G$. This completes the proof. $\square$

We close this section with the

PROOF OF THEOREM 5.5.16. By passing over to a Morita equivalent action we assume that τ is trivial and that G is equal to the vector group V.

Let $\Omega = \widehat{A}$ and let (x_n, s_n) be a sequence in Ω^S such that $h^\alpha(x_n, s_n)$ converges to $h^\alpha(x, s)$ in $\Omega \times_{S^\perp} \widehat{V}$. Then $x_n \to x$ in Ω and $S_{x_n} \to S_x$ in $\mathfrak{K}(V)$. By Lemma 5.5.18 we may assume that there exist elements $c_n \in GL(V)$ such that $c_n(S_x) = S_{x_n}$ for all $n \in \mathbb{N}$ and $c_n \to 1_V$ in $GL(V)$. Let $M = \{x_n; n \in \mathbb{N}\} \cup \{x\}$ and define $\tilde{h} : M \times S_x \to M \times \widehat{S}_x$ by $\tilde{h}(x_n, s) = h^\alpha(x_n, c_n(s)) \circ c_n$ and $\tilde{h}(x, s) = h^\alpha(x, s)$. Then $\tilde{h} = h^\beta$ if β denotes the action of S_x on A_M defined by the family $(\beta^y)_{y \in M}$ of actions of S_x on the fibers A_y given by $\beta_s^y(a(y)) = \alpha_{c_n(s)}^y(a(y))$ for all $y \in M$. Thus, we can use Theorem 5.5.13 in order to conclude that $\tilde{h}$ is open as a map onto its image. Similar arguments as used at the end of the previous proof then show that h^α is also open as a map onto its image. $\square$

CHAPTER 6

Applications and Examples

In this chapter we give some more applications of the methods and results which were developed and proved in the previous chapters. We also give examples which illustrate some subtleties of the theory.

6.1. Crossed products by σ-proper actions

In this section we want to investigate the structure of the primitive ideal spaces of twisted crossed products $A \rtimes_{\alpha,\tau} G$, if (A, G, α, τ) is a separable abelian twisted system such that A has continuous trace and the action of G on $\Omega = \widehat{A}$ is σ-proper. If, additionally, the symmetrizer map $\Sigma : \Omega \to \mathfrak{K}(G); x \mapsto \Sigma_x$ is continuous, then we will see that the primitive ideal space of $A \rtimes_{\alpha,\tau} G$ is always a σ-proper $\widehat{\widehat{G}}$-space. Recall that by Lemma 5.5.7 we know that $(A \rtimes_{\alpha,\tau} \Omega^\Sigma)\widehat{}$ is canonically G- and $\widehat{\widehat{G}}$-homeomorphic to $\mathrm{Prim}(A \rtimes_{\alpha,\tau} \Omega^S)$ via induction of representations. If we put $\mathcal{W} = (A \rtimes_{\alpha,\tau} \Omega^\Sigma)\widehat{}$, we may consider the maps:

$$\mathrm{ind} : \mathcal{W} \to \mathrm{Prim}(A \rtimes_{\alpha,\tau} G); (x, \rho \times V) \mapsto \ker(\mathrm{ind}_{\Sigma_x}^G (\rho \times V))$$

and

$$\mathrm{res} : \mathcal{W} \to \widehat{A}; (x, \rho \times V) \mapsto \rho.$$

Moreover, let $q : \widehat{A} \to \widehat{A}/G$ denote the quotient map and let $\mathrm{Res}^G : \mathrm{Prim}(A \rtimes_{\alpha,\tau} G) \to \widehat{A}/G$ be the canonical map defined by

$$\mathrm{Res}^G(\ker(\pi \times U)) = G(\rho) \Leftrightarrow \ker \pi = \ker G(\rho)$$

(see Proposition 1.4.13). Note that in the present situation $\widehat{A}/G$ is Hausdorff, and hence equal to the quasi-orbit space $\mathcal{Q}_G(\widehat{A})$. The following theorem generalizes [46, Corollary 2.1] and [19, Theorem 4] to the case of continuously varying stabilizers and symmetrizers and gives a very general version of [56, Theorem 6.3].

THEOREM 6.1.1. Let (A, G, α, τ) be a separable abelian twisted system such that G acts σ-properly on $\Omega = \widehat{A}$. Then $\mathrm{Prim}(A \rtimes_{\alpha,\tau} G)$ is Hausdorff if and only *if the symmetrizer map* $\Sigma : \widehat{A} \to \mathfrak{K}(G)$ *is continuous. Moreover, if this is true and if* $\mathcal{W} = (A \rtimes_{\alpha,\tau} \Omega^{\Sigma})\widehat{\ }$, *then the diagram*

$$
\begin{array}{ccc}
\mathcal{W} & \xrightarrow{\text{ind}} & \operatorname{Prim}(A \rtimes_{\alpha,\tau} G) \\
{\scriptstyle \text{res}}\downarrow & & \downarrow {\scriptstyle \text{Res}^G} \\
\widehat{A} & \xrightarrow[q]{} & \widehat{A}/G
\end{array}
$$

commutes and all maps in this diagram are quotient maps of σ-*proper* G- *or* $\widehat{\widehat{G}}$-*spaces, respectively: The horizontal arrows are quotient maps for* σ-*proper* G-*spaces and the vertical arrows are quotient maps for* σ-*proper* $\widehat{\widehat{G}}$-*spaces. In particular,* $\operatorname{Prim}(A \rtimes_{\alpha,\tau} G)$ *is a* σ-*proper* $\widehat{\widehat{G}}$-*space.*

For the proof we need

PROPOSITION 6.1.2. *Let* (Ω, R) *be an open* σ-*trivial regularization of the abelian twisted system* (A, G, α, τ) *and let* $\Lambda \subseteq \Omega$ *be the image of some continuous section* $\mathfrak{s} : \Omega/G \to \Omega$. *Let* $(\alpha^{\Lambda}, \tau^{\Lambda})$ *denote the subgroup action of* Λ^S *on* $A_{R^{-1}(\Lambda)} = A/\ker R^{-1}(\Lambda)$ *induced by* (α, τ) *(see Section 4.2), where* $S : \Omega \to \mathfrak{K}(G); x \mapsto S_x$ *denotes the stabilizer map. Moreover, let* $\widehat{\alpha}_{\Lambda}$ *denote the dual action of* $\widehat{\widehat{G}}$ *on* $A_{R^{-1}(\Lambda)} \rtimes_{\alpha^{\Lambda}, \tau^{\Lambda}} \Lambda^S$ *and let* $\widehat{\alpha}$ *be the usual dual action of* $\widehat{\widehat{G}}$ *on* $A \rtimes_{\alpha,\tau} G$. *Then* $\widehat{\alpha}_{\Lambda}$ *is Morita equivalent to* $\widehat{\alpha}$.

PROOF. It is a direct consequence of [**13**, Theorem 2] that $A_{R^{-1}} \rtimes_{\alpha^{\Lambda}, \tau^{\Lambda}} \Lambda^S$ is Morita equivalent to $A \rtimes_{\alpha,\tau} G$ (see also Proposition 5.1.7). In order to prove the desired Morita equivalence of the dual actions, we have to look more closely at the construction of the imprimitivity bimodule. For this we use Proposition 5.1.2 in order to write $\Omega = \Lambda \times_S G$. Using Proposition 4.2.3 we conclude easily that there is a canonical surjective *-homomorphism

$$
\Phi : C^*(\Lambda^S, A, \alpha, \tau) \to A_{R^{-1}(\Lambda)} \rtimes_{\alpha^{\Lambda}, \tau^{\Lambda}} \Lambda^S
$$

with kernel

$$
\ker \Phi = \bigcap \{ \ker(x, \rho \times V) \in C^*(\Lambda^S, A, \alpha, \tau)\widehat{\ }; \ker \rho \supseteq \ker R^{-1}(\{x\}) \}.
$$

Moreover, in view of Lemma 1.4.10 we see that there is a canonical G-equivariant *-homomorphism Ψ_R from $C_0(\Lambda \times_S G, A)$ onto A, which gives rise to a canonical surjective *-homomorphism

$$
\Psi : C_0(\Lambda \times_S G, A) \rtimes_{\tilde{\alpha}, \tilde{\tau}} G \to A \rtimes_{\alpha,\tau} G,
$$

where $(\tilde{\alpha}, \tilde{\tau})$ denotes the diagonal twisted action of G on $C_0(\Lambda \times_S G, A)$. It is trivially seen that both Φ and Ψ are $\widehat{\widehat{G}}$-equivariant with respect to the dual actions.

Now let $\mathcal{X}$ be the $C_0(\Lambda \times_S G, A) \rtimes_{\tilde{\alpha}, \tilde{\tau}} G - C^*(\Lambda^L, A, \alpha, \tau)$ imprimitivity bimodule as constructed in Proposition 4.4.3. Recall that $\mathcal{X}$ is obtained as a completion of the $C_c(G, C_c(\Lambda \times_S G, A), \tilde{\tau}) - C_c(\Lambda^S, A, \tau)$ imprimitivity bimodule $\mathcal{X}_0 = C_c(\Lambda \times G, A, \tau)$. The proof of [**13**, Theorem 2] shows that $\ker \Psi$ is exactly the ideal induced from $\ker \Phi$ via $\mathcal{X}$. Thus it is enough to show that there exists an action, say $\widehat{u}$, of $\widehat{\tilde{G}}$ on $\mathcal{X}$ such that $(\mathcal{X}, \widehat{u})$ becomes a $\widehat{\tilde{\alpha}} - \widehat{\alpha}_\Lambda$ imprimitivity pair, where, for the moment, $\widehat{\alpha}_\Lambda$ denotes the dual action of $\widehat{\tilde{G}}$ on $C^*(\Lambda^S, A, \alpha, \tau)$. But if we define $\widehat{u}$ (as usual) by

$$\widehat{u}_\chi \xi(x, s) = \overline{\chi(s)} \xi(x, s)$$

for all $\xi \in \mathcal{X}_0$, then it follows directly from the formulas for the actions and inner products on $\mathcal{X}_0$ as given in the remarks preceding Proposition 4.4.3 that $\widehat{u}$ does the job. $\square$

PROOF OF THEOREM 6.1.1. Suppose that $\Sigma : \widehat{A} \to \mathfrak{K}(G)$ is not continuous. By the compactness of $\mathfrak{K}(G)$ and Corollary 5.4.6 there exists a convergent sequence $\rho_n \to \rho$ in $\widehat{A}$ such that $\Sigma_{\rho_n} \to \Sigma_0 \subsetneq \Sigma_\rho$. Let $V : \Sigma_\rho \to \mathcal{U}(\mathcal{H}_\rho)$ be a unitary representation extending ρ to the representation $\rho \times V$ of $A \rtimes_{\alpha, \tau} \Sigma_\rho$, and let $W = V|_{\Sigma_0}$. Since

$$(\Sigma_{\rho_n}, \operatorname{ind}_{N_\tau}^{\Sigma_{\rho_n}} \rho_n) \to (\Sigma_0, \operatorname{ind}_{N_\tau}^{\Sigma_0} \rho)$$

in $\mathcal{S}(\mathfrak{K}(G)^{\mathrm{Id}}, A, \alpha, \tau)$ by the continuity of induction (see Proposition 4.1.3), and since $\rho \times W$ is weakly contained in $\operatorname{ind}_{N_\tau}^{\Sigma_0} \rho = \lambda_{\Sigma_0} \otimes (\rho \times W)$ (see Proposition 1.4.9), we conclude from Propositions 1.2.1 and 1.2.3 that, by passing to a subsequence if necessary, there exist extensions $\rho_n \times V^n$ of ρ_n to $A \rtimes_{\alpha, \tau} \Sigma_{\rho_n}$ such that

$$(\Sigma_{\rho_n}, \rho_n \times V^n) \to (\Sigma_0, \rho \times W)$$

in $C^*(\mathfrak{K}(G)^{\mathrm{Id}}, A, \alpha, \tau)^\frown$. By Theorem 3.1.11 we know that $P_n = \ker(\operatorname{ind}_{\Sigma_{\rho_n}}^G (\rho_n \times V^n))$ is a primitive ideal of $A \rtimes_{\alpha, \tau} G$ for all $n \in \mathbb{N}$. By the continuity of induction and Proposition 1.2.1 we conclude that $(P_n)_{n \in \mathbb{N}}$ converges to each primitive ideal of $A \rtimes_{\alpha, \tau} G$ which contains $\ker(\operatorname{ind}_{\Sigma_0}^G (\rho \times W))$. But applying Green's theorem (Theorem 1.4.12) to the quotient system $A_{G(\rho)} \rtimes_{\alpha, \tau} G$, we see that $\operatorname{Prim}(A_{G(\rho)} \rtimes_{\alpha, \tau} G)$ is homeomorphic to $\operatorname{Prim}(A_\rho \rtimes_{\alpha, \tau} S_\rho)$ which in turn is homeomorphic to $(A_\rho \rtimes_{\alpha, \tau} \Sigma_\rho)^\frown$, where all homeomorphism are given by induction of representations. Moreover, $(A_\rho \rtimes_{\alpha, \tau} \Sigma_\rho)^\frown$ is homeomorphic to $\widehat{\Sigma_\rho / N_\tau}$ via the map $\chi \to \rho \times \chi V$ and we have $\operatorname{ind}_{\Sigma_0}^{\Sigma_\rho}(\rho \times W) = \lambda_{\Sigma_\rho / \Sigma_0} \otimes (\rho \times V)$ by Proposition 1.4.9, from which follows that $\operatorname{ind}_{\Sigma_0}^{\Sigma_\rho}(\rho \times W)$ is weakly equivalent to the set $\{\rho \times \mu V \in (A_\rho \rtimes_{\alpha, \tau} \Sigma_\rho)^\frown; \mu \in \widehat{\Sigma_\rho / \Sigma_0}\}$. Since by the theorem of induction in stages $\operatorname{ind}_{\Sigma_0}^G (\rho \times W) = \operatorname{ind}_{\Sigma_\rho}^G (\operatorname{ind}_{\Sigma_0}^{\Sigma_\rho}(\rho \times W))$, it follows now that the set of primitive ideals of $A \rtimes_{\alpha, \tau} G$ which contain the kernel of $\operatorname{ind}_{\Sigma_0}^G (\rho \times W)$ is homeomorphic to $\widehat{\Sigma_\rho / \Sigma_0}$, which has more than one element by assumption. We conclude that $\operatorname{Prim}(A \rtimes_{\alpha, \tau} G)$ is not Hausdorff.

Assume now that $\Sigma : \Omega = \widehat{A} \to \mathfrak{K}(G)$ is continuous. It is an easy consequence of Proposition 1.4.8 that the diagram commutes. It follows from Proposition

5.2.9 that res is the quotient map of a σ-proper $\widehat{\widehat{G}}$-space, and it follows from Proposition 5.1.8 together with Lemmas 1.4.10 and 5.5.10 that $\mathcal{W}$ is a σ-proper G-space with quotient map ind. Thus, for the proof of the theorem we only have to show that $\mathrm{Prim}(A \rtimes_{\alpha,\tau} G)$ is a σ-proper $\widehat{\widehat{G}}$-space with quotient map Res^G.

We know from Proposition 3.3.1 that Res^G induces a homeomorphism between the quasi-orbit space $\mathcal{Q}_{\widehat{\widehat{G}}}(\mathrm{Prim}(A \rtimes_{\alpha,\tau} G))$ and $\widehat{A}/G$. But by the Mackey-Green machine (see Theorem 1.4.12 and Proposition 3.1.13) it follows that $\mathrm{Prim}(A_{G(\rho)} \rtimes_{\alpha,\tau} G)$ consists of a single $\widehat{\widehat{G}}$-orbit, which implies that the $\widehat{\widehat{G}}$-orbits are closed in $\mathrm{Prim}(A \rtimes_{\alpha,\tau} G)$ since $G(\rho)$ is closed in $\widehat{A}$ by assumption. Thus we conclude that the quasi-orbit space of $\mathrm{Prim}(A \rtimes_{\alpha,\tau} G)$ is equal to the orbit space and that Res^G may be viewed as the corresponding quotient map. In order to show that $\mathrm{Prim}(A \rtimes_{\alpha,\tau} G)$ is a σ-proper $\widehat{\widehat{G}}$-space, let $(G(\rho_n))_{n \in \mathbb{N}}$ be a sequence in $\widehat{A}/G$ which converges to $G(\rho)$ for some $\rho \in \widehat{A}$. By passing to a subsequence if necessary, we may assume by Proposition 5.1.3 that $q^{-1}(M)$ is a σ-trivial G-space if M denotes the closure of $\{G(\rho_n); n \in \mathbb{N}\} \cup \{G(\rho)\}$. Let Λ be the image of some continuous section $\mathfrak{s} : M \to M/G$. Then it follows from Proposition 6.1.2 that $\mathrm{Prim}(A_M \rtimes_{\alpha,\tau} G)$ is $\widehat{\widehat{G}}$-homeomorphic to $\mathrm{Prim}(A_\Lambda \rtimes_{\alpha^\Lambda,\tau^\Lambda} \Lambda^S)$. But $\mathrm{Prim}(A_\Lambda \rtimes_{\alpha^\Lambda,\tau^\Lambda} \Lambda^S)$ is $\widehat{\widehat{G}}$-homeomorphic to $(A_\Lambda \rtimes_{\alpha^\Lambda,\tau^\Lambda} \Lambda^\Sigma)\widehat{}$, which follows from Lemma 5.5.7. But $(A_\Lambda \rtimes_{\alpha^\Lambda,\tau^\Lambda} \Lambda^\Sigma)\widehat{}$ is a σ-proper $\widehat{\widehat{G}}$-space by Theorem 5.2.9. Thus we conclude that $\mathrm{Prim}(A_M \rtimes_{\alpha,\tau} G)$ is a σ-proper $\widehat{\widehat{G}}$-space. By passing to another subsequence, say $(G(\rho_{n_m}))_{m \in \mathbb{N}}$, of $(G(\rho_n))_{n \in \mathbb{N}}$ we conclude from Proposition 5.1.3 that $(\mathrm{Res}^G)^{-1}(\tilde{M})$ is a σ-trivial $\widehat{\widehat{G}}$-space, where $\tilde{M} = \{G(\rho_{n_m}); m \in \mathbb{N}\} \cup \{G(\rho)\}$. But this implies that $\mathrm{Prim}(A \rtimes_{\alpha,\tau} G)$ is Hausdorff and, again using Proposition 5.1.3, that $\mathrm{Prim}(A \rtimes_{\alpha,\tau} G)$ is a σ-proper $\widehat{\widehat{G}}$-space. $\square$

One special case of Theorem 6.1.1, which is particularly interesting, is the case where G acts trivially on $\widehat{A}$. Here we get the following corollary.

COROLLARY 6.1.3. *Suppose that* (A, G, α, τ) *is a separable abelian twisted system such that* A *has continuous trace and* G *acts trivially on* $\widehat{A}$. *Then* $\mathrm{Prim}(A \rtimes_{\alpha,\tau} G)$ *is Hausdorff if and only if the symmetrizer map* $\Sigma : \widehat{A} \to \mathfrak{K}(G)$ *is continuous. If this is true, then* $\mathrm{Prim}(A \rtimes_{\alpha,\tau} G)$ *is a* σ*-proper* $\widehat{\widehat{G}}$*-space.*

In the remainder of this section we want to show that the results and methods developed in this work may be used to give a new proof of Rosenberg's fundamental result that pointwise unitary actions of compactly generated abelian groups on continuous trace C^*-algebras are automatically locally unitary (see [64, Corollary 2.2] – for the twisted situation see [46, Proposition 1.4]).

Recall that if (A, G, α, τ) is a twisted covariant system, then (α, τ) is called locally unitary if G acts trivially on $\widehat{A}$ and for each $\rho \in \widehat{A}$ there exists an open neighborhood W of ρ such that (α^W, τ^W) is a unitary twisted action of G on

A_W. By Proposition 2.1.3 we know that a twisted action (α, τ) of G on A is unitary if and only if it is Morita equivalent to a trivial action of $\tilde{G} = G/N_\tau$ on some C^*-algebra B (which then trivially implies that it is Morita equivalent to the trivial action of $\tilde{G}$ on A). By localizing, this easily implies that the property of being locally unitary is inherited to Morita equivalent twisted actions. We use this to prove

THEOREM 6.1.4 (ROSENBERG). *Suppose that (A, G, α, τ) is a separable twisted abelian system such that A has continuous trace and $\tilde{G} = G/N_\tau$ is compactly generated. Assume further that (α, τ) is pointwise unitary. Then (α, τ) is locally unitary.*

For the proof we need the following result.

PROPOSITION 6.1.5. *Let (A, G, α) be an abelian system such that $\mathrm{Prim}(A)$ is a locally trivial G-bundle. Then the dual action $\hat{\alpha}$ is locally unitary.*

PROOF. By localizing we may assume that $\mathrm{Prim}(A)$ is a trivial G-bundle. So let Λ be the image of a continuous section $\mathfrak{s} : \mathrm{Prim}(A)/G \to \mathrm{Prim}(A)$ and let $A_\Lambda = A/\ker \Lambda$. Then it follows from Proposition 6.1.2 that the trivial action of $\hat{G}$ on A_Λ is Morita equivalent to (α, τ), which implies that (α, τ) is unitary by Proposition 2.1.3. $\square$

PROOF OF THEOREM 6.1.4. Identifying $A \rtimes_{\alpha,\tau} G$ with $A \rtimes_{\alpha,\tau} \Omega^G$, where G is identified with the constant map $\Omega = \hat{A} \to \mathfrak{K}(G); x \mapsto G$, it follows from Theorem 5.2.9 that $(A \rtimes_{\alpha,\tau} G)\hat{\ }$ is a proper $\hat{\tilde{G}}$-space. Since $\tilde{G}$ is compactly generated it follows that $\hat{\tilde{G}}$ is a Lie group. Thus we can apply Palais's slice theorem [48, §4.1] in order to see that $(A \rtimes_{\alpha,\tau} G)\hat{\ }$ is a locally trivial $\hat{\tilde{G}}$-bundle. But Proposition 6.1.5 then implies that the double dual action $\hat{\hat{\alpha}}$ of $\tilde{G}$ on $(A \rtimes_{\alpha,\tau} G) \rtimes_{\hat{\alpha}} \hat{\tilde{G}}$ is locally unitary. Since $\hat{\hat{\alpha}}$ is Morita equivalent to (α, τ) by Proposition 2.2.1, we conclude that (α, τ) is locally unitary, too. $\square$

Again we should mention that, unlike [64, Corollary 2.2], our proof does not give any information about actions of non-abelian groups, so that Rosenberg's original result is more general.

6.2. Actions of $\mathbb{R}$, $\mathbb{Z}$ and $\mathbb{T}$

We now reduce our attention to twisted covariant systems (A, G, α, τ) such A has continuous trace and $\tilde{G} = G/N_\tau$ is isomorphic to either the reals $\mathbb{R}$, the integers $\mathbb{Z}$, the one-dimensional torus $\mathbb{T}$, or a finite cyclic group $\mathbb{Z}/n\mathbb{Z}$ for some $n \in \mathbb{N}$. Note that in all these situations the Mackey obstructions of (A, G, α, τ) vanish since every multiplier on a group of the form $\mathbb{R}$, $\mathbb{Z}$, $\mathbb{T}$ or $\mathbb{Z}/n\mathbb{Z}$ is trivial. We will now see that for actions of such groups the continuous trace problem for the crossed product has a solution similar to the solution of the continuous trace problem for abelian transformation groups given by Williams in [**70**, Theorem 5.1].

THEOREM 6.2.1. *Let (A, G, α, τ) be a separable twisted covariant system such that A has continuous trace and such that $\tilde{G} = G/N_\tau$ is isomorphic to either $\mathbb{R}$, $\mathbb{Z}$, $\mathbb{T}$ or $\mathbb{Z}/n\mathbb{Z}$ for some $n \in \mathbb{N}$. Then $A \rtimes_{\alpha,\tau} G$ has continuous trace if and only if G acts σ-properly on $\widehat{A} = \Omega$. In particular, if this is true, then the stabilizer map $S : \Omega \to \mathfrak{K}(G)$ is continuous and we get the following commutative diagram*

$$
\begin{array}{ccc}
(A \rtimes_{\alpha,\tau} \Omega^S)\widehat{\;} & \xrightarrow{\ \text{ind}\ } & (A \rtimes_{\alpha,\tau} G)\widehat{\;} \\
{\scriptstyle\text{res}}\downarrow & & \downarrow{\scriptstyle\text{Res}^G} \\
\widehat{A} & \xrightarrow[\ q\ ]{} & \widehat{A}/G
\end{array}
$$

in which each map is a quotient map of a σ-proper G- or $\widehat{\tilde{G}}$-space. Moreover, we have the relation

$$
\text{res}^*(\delta(A)) = \text{ind}^*(\delta(A \rtimes_{\alpha,\tau} G))
$$

for the Dixmier-Douady classes $\delta(A)$ and $\delta(A \rtimes_{\alpha,\tau} G)$ of A and $A \rtimes_{\alpha,\tau} G$, respectively.

PROOF. Since all assertions of the theorem are invariant under passing over to a Morita equivalent action we assume that τ is trivial. Assume first that G acts σ-properly on $\widehat{A}$. Then it follows from Lemma 5.5.10 that G acts σ-properly on $(A \rtimes_\alpha \Omega^S)\widehat{\;}$, from which then follows by Theorem 5.3.3 that $A \rtimes_\alpha G$ has continuous trace.

Assume now that $A \rtimes_\alpha G$ has continuous trace. Then it follows from Theorem 5.3.3 that the stabilizer map $S : \Omega \to \mathfrak{K}(G)$ is continuous. If $G = \mathbb{T}$ or $\mathbb{Z}/n\mathbb{Z}$, then it follows directly that G acts σ-properly on $\widehat{A}$. If $G = \mathbb{Z}$, then it follows from Theorem 3.3.3 and the continuity of the map $H \mapsto H^\perp$ that the dual system $(A \rtimes_\alpha \mathbb{Z}, \widehat{\mathbb{Z}}, \widehat{\alpha})$ has a continuous stabilizer map, too. Since $\widehat{\mathbb{Z}} \cong \mathbb{T}$ is compact, this implies that $(A \rtimes_\alpha \mathbb{Z})\widehat{\;}$ is a σ-proper $\widehat{\mathbb{Z}}$-space, from which then follows by Theorem 6.1.1 that $(A \rtimes_\alpha \mathbb{Z}) \rtimes_{\widehat{\alpha}} \widehat{\mathbb{Z}}$ is a σ-proper $\mathbb{Z}$-space with respect to the double dual action $\widehat{\widehat{\alpha}}$. But this implies that $\widehat{A}$ is a σ-proper G-space, since α is Morita equivalent to $\widehat{\widehat{\alpha}}$.

Let us now assume that $G = \mathbb{R}$. It follows from [**54**, Lemma 4.15] that $\mathfrak{K}(\mathbb{R})$ is homeomorphic to $[0, \infty]$ via the map

$$
c(t) = \begin{cases} \{0\} & \text{for } t = \infty, \\ \mathbb{R} & \text{for } t = 0, \text{ and} \\ t\mathbb{Z} & \text{for } 0 < t < \infty. \end{cases}
$$

We assume for the moment that $c^{-1}(S_x) \leq 1$ for each $x \in \Omega$. Then the quotient map $q : \Omega \times \mathbb{R} \to \Omega \times_S \mathbb{R}$ maps $\Omega \times [0, 1]$ onto all of $\Omega \times_S \mathbb{R}$. Thus $\Omega \times_S \mathbb{R}$

is compact and $p : \Omega \times_S \mathbb{R} \to \Omega \times \Omega; (x, s) \to (x, sx)$ is proper. Thus $\mathbb{R}$ acts σ-properly on Ω.

Suppose now that $c^{-1}(S_x) \geq 1$ for all $x \in \Omega$. Let us identify $\widehat{\mathbb{R}}$ with $\mathbb{R}$ via the canonical map $s \to \chi_s$ with $\chi_s(t) = e^{i2\pi st}$. Then it follows from trivial computations that

$$c^{-1}(S_x^\perp) = \frac{1}{c^{-1}(S_x)} \leq 1$$

for all $x \in \Omega$. Thus it follows from Theorem 3.3.3 that the stabilizers of the dual system $(A \rtimes_\alpha \mathbb{R}, \mathbb{R}, \widehat{\alpha})$ vary continuously and have image less than 1 under the map c^{-1}. We conclude from this that $(A \rtimes_\alpha \mathbb{R})\widehat{}$ is a σ-proper $\mathbb{R}$-space with respect to the dual action $\widehat{\alpha}$. But then Theorem 6.1.1 and the Morita equivalence between α and $\widehat{\alpha}$ imply that $\widehat{A}$ is a σ-proper $\mathbb{R}$-space, too. The general case follows easily by writing Ω as the union of the two closed G-invariant subsets $S^{-1}(c([0, 1]))$ and $S^{-1}(c([1, \infty]))$.

All other assertions of the theorem follow from Theorem 5.3.3 and Theorem 6.1.1. $\square$

6.3. Examples and applications to certain group C^*-algebras

The following example serves as a counter example for many things one could expect while investigating crossed products with continuous trace. It shows that the quasi-orbit space $\mathcal{Q}_G(\widehat{A})$ need not be Hausdorff in order that $A \rtimes_\alpha G$ be a continuous-trace C^*-algebra. It also serves as an example for a system which has continuously varying maximally pointwise unitary subgroups in the sense of Definition 5.4.1 but there exists no continuous choice of maximally pointwise unitary subgroups for the system. Since the group G in the example is $\mathbb{R}^2$, it also shows that the results for actions of $\mathbb{R}$, $\mathbb{Z}$ and $\mathbb{T}$ as given in Theorem 6.2.1 are not valid if we consider groups of higher dimension.

Example 6.3.1. Let θ be the action of $G = \mathbb{R}^2$ on the real line $\mathbb{R}$ defined as follows:

$$\theta_{(s,t)}(r) = \begin{cases} e^s r & \text{for } r > 0 \\ 0 & \text{for } r = 0 \\ e^t r & \text{for } r < 0, \end{cases}$$

and let ω be the multiplier on $\mathbb{R}^2$ defined by

$$\omega((s, t), (s', t')) = e^{i2\pi st'}$$

for $(s, t), (s', t') \in \mathbb{R}^2$. Let $L : \mathbb{R}^2 \to \mathcal{U}(L^2(\mathbb{R}^2))$ denote the regular ω-representation of $\mathbb{R}^2$ given by

$$(L_{(s,t)}\xi)(s', t') = e^{i2\pi s(t'-t)}\xi(s' - s, t' - t), \quad \xi \in L^2(\mathbb{R}^2).$$

Define $\mathcal{K} = \mathcal{K}(L^2(\mathbb{R}^2))$ and let $\alpha = \theta \otimes \operatorname{Ad} L$ be the the diagonal action of $\mathbb{R}^2$ on $A = C_0(\mathbb{R}, \mathcal{K})$, i.e.,

$$(\alpha_{(s,t)}(f))(r) = L_{(s,t)} f(\theta_{(-s,-t)}(r)) L^*_{(s,t)}$$

for all $f \in A$. There are exactly three orbits in $\widehat{A} \cong \mathbb{R}$ under the action α, namely the orbits $(-\infty, 0)$, $\{0\}$ and $(0, \infty)$, and it is clear that $\{0\}$ is contained in the closure of $(-\infty, 0)$ as well as in the closure of $(0, \infty)$. Thus we see that $\widehat{A}/G = Q_{\mathbb{R}^2}(\widehat{A})$ is not Hausdorff.

The stability groups of the system are given by

$$S_{\rho_r} = \begin{cases} \mathbb{R} \times \{0\} & \text{if } r < 0 \\ \mathbb{R}^2 & \text{if } r = 0, \text{ and} \\ \{0\} \times \mathbb{R} & \text{if } r > 0, \end{cases}$$

where for each $r \in \mathbb{R}$, ρ_r denotes the element of $\widehat{A}$ corresponding to r via evaluation at r. Since $\mathbb{R}$ only carries trivial multipliers, it follows that the only choice for a maximally ρ_r-unitary subgroup of $\mathbb{R}^2$ is the group $H_{\rho_r} = \mathbb{R} \times \{0\}$ if $r < 0$ and $H_{\rho_r} = \{0\} \times \mathbb{R}$ if $r > 0$. If $r = 0$, then both, $\mathbb{R} \times \{0\}$ and $\{0\} \times \mathbb{R}$ are maximally ρ_r-unitary subgroups of $\mathbb{R}^2$ for $(A, \mathbb{R}^2, \alpha)$, since $[\omega]$ is the Mackey obstruction at ρ_0 and both groups are maximally ω-trivial. Thus it turns out that the system $(A, \mathbb{R}^2, \alpha)$ has continuously varying maximally pointwise unitary subgroups in the sense of Definition 5.4.1, but there exists no continuous choice of maximally pointwise unitary subgroups for $(A, \mathbb{R}^2, \alpha)$.

Since $Q_{\mathbb{R}^2}(\widehat{A})$ is not Hausdorff and the stabilizer map is not continuous, there is no result in this work which can be applied directly to $(A, \mathbb{R}^2, \alpha)$ in order to see that $A \rtimes_\alpha \mathbb{R}^2$ has continuous trace. However, it follows easily from Proposition 5.1.4 that $A \rtimes_\alpha \mathbb{R}^2$ has continuous trace if we can show that this is true for the quotients $A_{(-\infty,0]} \rtimes_\alpha \mathbb{R}^2$ and $A_{[0,\infty)} \rtimes_\alpha \mathbb{R}^2$, where $A_{(-\infty,0]} = C_0((-\infty, 0], \mathcal{K})$ and $A_{[0,\infty)} = C_0([0, \infty), \mathcal{K})$. Since the situation is completely symmetric, it is enough to show that $A_{[0,\infty)} \rtimes_\alpha \mathbb{R}^2$ has continuous trace. If we put $H = \{0\} \times \mathbb{R}$, then H is a constant choice of maximally pointwise unitary subgroups for $(A_{[0,\infty)}, \mathbb{R}^2, \alpha)$, and by Theorem 5.3.2 it suffices to show that the natural action of $\mathbb{R}^2$ on $(A \rtimes_\alpha H)\widehat{\ }$ factors through a proper action of $\mathbb{R} \cong \mathbb{R}^2/H$. For this let $V = L|_H$. Then we see that the action of $\mathbb{R} \cong H$ on $A_{[0,\infty)} = C_0([0, \infty), \mathcal{K})$ is implemented by the strictly continuous homomorphism

$$v : \mathbb{R} \to \mathcal{U}(A_{[0,\infty)}); (v_t f)(r) = V_t f(r).$$

It follows that $(A \rtimes_\alpha H)\widehat{\ }$ is homeomorphic to $[0, \infty) \times \mathbb{R}$ via the map

$$(r, l) \mapsto \rho_r \times \chi_l(\rho_r \circ v),$$

where for each $l \in \mathbb{R}$, χ_l denotes the character on $\mathbb{R} \cong H$ given by $\chi_l(t) = e^{2\pi i l t}$. Now let $s \in \mathbb{R} \cong \mathbb{R}^2/H$ and let $\rho_r \times \chi_l(\rho_r \circ v)$ be a representation of $A \rtimes_\alpha H$ as above. Then, using Lemma 3.1.9, we compute for every $g \in C_c(H, A_{[0,\infty)})$:

$$(\rho_r \times \chi_l(\rho_r \circ v)) \circ \gamma^H_{(s,0)}(g) = \int_H \rho_r(\alpha_{(s,0)}(g(t)))\chi_l(t)(\rho_r \circ v)(t)\, dt$$

$$= \int_H (\alpha_{(s,0)}(g(t)))(r)\chi_l(t)V_t\, dt = \int_H L_{(s,0)}g(t)(e^{-s}r)L^*_{(s,0)}L_{(0,t)}\chi_l(t)\, dt$$

$$= L_{(s,0)} \left(\int_H g(t)(e^{-s}r)h_\omega(s,0)(0,t)\chi_l(t)V_t \, dt \right) L^*_{(s,0)}$$

$$= L_{(s,0)} \left(\rho_{e^{-s}r} \times h_\omega(s,0)\chi_l(\rho_{e^{-s}r} \circ v)(g) \right) L^*_{(s,0)}.$$

Since $h_\omega(s,0)(0,t) = e^{2\pi i s t} = \chi_s(t)$, we see that $(\rho_r \times \chi_l(\rho_r \circ V)) \circ \gamma^H_{(s,0)}$ is unitarily equivalent to $\rho_{e^{-s}r} \times \chi_{s+l}(\rho_{e^{-s}r} \circ v)$. Thus, the action of $\mathbb{R} \cong \mathbb{R}^2/H$ on $[0,\infty) \times \mathbb{R} \cong (A_{[0,\infty)} \rtimes_\alpha H)\widehat{\ }$ is given by $s \cdot (r,l) = (e^{-s}r, s+l)$. If we define the map $\phi : [0,\infty) \times \mathbb{R} \to [0,\infty) \times \mathbb{R}$ by $\phi(r,l) = (e^l r, l)$, then ϕ is a homeomorphism which carries this action to the action which is given by translation of the second variable. Thus $(A \rtimes_\alpha H)\widehat{\ }$ is a trivial $\mathbb{R}$-space and $A \rtimes_\alpha \mathbb{R}^2$ has continuous trace. Let us finally mention that the crossed product $A \rtimes_\alpha \mathbb{R}^2$ is even isomorphic to A. Using [9, Theorem 10.8.8], this follows from the fact that $(A \rtimes_\alpha \mathbb{R}^2)\widehat{\ } \cong \mathbb{R}$, which follows quite easily from continuity of restriction and induction, and the triviality of the Dixmier-Douady class of $A \rtimes_\alpha \mathbb{R}^2$ (since $H^3(\mathbb{R}, \mathbb{Z})$ is trivial).

The next example shows that some of the obscurities in the above example can also happen if G is finite.

Example 6.3.2. Let $\mathbb{Z}_2 = \{1, -1\}$ and let $G = \mathbb{Z}_2 \times \mathbb{Z}_2$. We put $\Omega = \{(r, z) \in \mathbb{R} \times \mathbb{C}; |r| = |z|\}$ and we define an action θ of G on Ω by

$$\theta_{(j,l)}(r,z) = \begin{cases} (r, lz) & \text{if } r \geq 0 \\ (r, jz) & \text{if } r \leq 0, \end{cases}$$

for all $(j,l) \in G$. Let ω be the multiplier on G given by $\omega((j,l),(j',l')) = jl'$ and let $L : G \to \mathcal{U}(l^2(G))$ be the left regular ω-representation given by

$$L_{(j,l)}f(j',l') = jl^{-1}l' f(j^{-1}j', l^{-1}l')$$

for all $f \in l^2(G) \cong \mathbb{C}^4$. As in the previous example we define an action α of G on $A = C_0(\Omega, M_4(\mathbb{C}))$ by

$$(\alpha_{(j,l)}(g))(r,z) = L_{(j,l)}g(\theta_{(j^{-1},l^{-1})}(r,z))L^*_{(j,l)},$$

for $(j,l) \in G$ and $(r,z) \in \Omega$. If we identify $\widehat{A}$ with Ω via $(r,z) \to \rho_{(r,z)}$, where $\rho_{(r,z)}$ denotes evaluation at (r,z), then we have the following stabilizers for (A, G, α):

$$S_{\rho_{(r,z)}} = \begin{cases} \mathbb{Z}_2 \times \{1\} & \text{if } r > 0, \\ G & \text{if } r = 0, \text{ and} \\ \{1\} \times \mathbb{Z}_2 & \text{if } r < 0. \end{cases}$$

Thus the stabilizer map is not continuous at $(0,0)$. Moreover, if $r \neq 0$, then the only maximally $\rho_{(r,z)}$-unitary subgroup of G is the stabilizer $S_{\rho_{(r,z)}}$. If $r = 0$ (and hence also $z = 0$), then both groups, $\mathbb{Z}_2 \times \{1\}$ and $\{1\} \times \mathbb{Z}_2$ are maximally $\rho_{(0,0)}$-unitary subgroups of G, since ω is the Mackey obstruction of (A, G, α) at $\rho_{(0,0)}$. It follows that (A, G, α) has continuously varying maximally pointwise unitary subgroups in the sense of Definition 5.4.1, but there exists no continuous choice of maximally pointwise unitary subgroups for (A, G, α). However, $A \rtimes_\alpha G$

has continuous trace by Corollary 5.4.7. In fact, $A \rtimes_\alpha G$ is Morita equivalent to A, since it is easily seen that $(A \rtimes_\alpha G)\widehat{} \cong \widehat{A}$ and $H^3(\widehat{A}, \mathbb{Z})$ is trivial.

In what follows we want to show how our results can be used in the investigation of the structure of group C^*-algebras for certain separable locally compact groups. Let us first assume that G is two-step solvable, i.e. there exists a closed normal subgroup N of G such that N and G/N are abelian.

In this case we can write $C^*(G)$ as the twisted crossed product $C_0(\widehat{N}) \rtimes_{\gamma^N, \tau^N} G$. Thus, since $C_0(\widehat{N})$ is commutative and, therefore, has continuous trace, we can use our results for the investigation of the structure of $C^*(G)$. For instance, if G is type I, then it is always interesting to find a composition sequence of ideals $(I_\nu)_{0 \le \nu \le \sigma}$ for $C^*(G)$ satisfying

(1) $I_0 = \{0\}$ and $I_\sigma = C^*(G)$.
(2) $I_\nu \subseteq I_{\nu+1}$ for all ν, and if ν is a limit ordinal, then $I_\nu = \overline{\bigcup_{\mu < \nu} I_\mu}$.
(3) $I_{\nu+1}/I_\nu$ has continuous trace for all ordinals $\nu \le \sigma$.

(for the existence of such composition sequence see [**9**, 4.5.6]). As one way to find such composition sequence of ideals we want to suggest the following method: Find a sequence $(W_\nu)_{0 \le \nu \le \sigma}$ of open G-invariant subsets of $\widehat{N}$ satisfying

(1) $W_0 = \emptyset$ and $W_\sigma = \widehat{N}$.
(2) $W_\nu \subseteq W_{\nu+1}$ for all ordinals ν, and if ν is a limit ordinal, then $W_\nu = \bigcup_{\mu < \nu} W_\mu$.
(3) Each system $C_0(W_{\nu+1} \setminus W_\nu) \rtimes_{\alpha, \tau} G$ has continuous trace.

If $(W_\nu)_{0 \le \nu \le \sigma}$ is such a sequence of open G-invariant subsets of $\widehat{N}$, then we put $I_\nu = C_0(W_\nu) \rtimes_{\gamma^N, \tau^N} G$ to obtain a composition sequence for $C^*(G)$.

Of course, we suggest using the results of the previous chapters in order to find such sequence of open sets in $\widehat{N}$. For this it is necessary to compute the Mackey obstructions of the system. This can be done very easily as follows: Let $c : \tilde{G} = G/N \to G; \tilde{s} \mapsto c_{\tilde{s}}$ be any Borel section with $c_{\tilde{e}} = 1$. Let $\chi \in \widehat{N}$ and let S_χ denote the stabilizer of χ. If we define

$$\chi'(s) = \chi(c_{\tilde{s}}^{-1} s) \quad \text{for all } s \in S_\chi,$$

then χ' becomes an ω_χ-extension of χ, where ω_χ denotes the multiplier on $\tilde{S}_\chi = S_\chi/N$ given by

$$\omega_\chi(\tilde{s}, \tilde{t}) = \chi(c_{\tilde{s}}^{-1} c_{\tilde{t}}^{-1} c_{\tilde{s}t}), \quad \tilde{s}, \tilde{t} \in \tilde{S}_\chi.$$

Thus, the Mackey obstruction of $(C_0(\widehat{N}), G, \gamma^N, \tau^N)$ at χ is equal to the class $[\omega_\chi]$ in $H^2(\tilde{S}_\chi, \mathbb{T})$. If K_χ denotes the group kernel of χ in N, then the symmetrizer Σ_χ of χ is just the pull-back of the center of S_χ/K_χ in G. Of course, in order to find maximally χ-unitary subgroups of G, one has to look more closely at the multipliers ω_χ.

Our method works especially well if G is two-step nilpotent, which means that $G/Z(G)$ is abelian, where $Z(G)$ denotes the center of G. If we put $N = Z(G)$ in the procedure above, then all stabilizers are equal to G. If, moreover, G/N

is a compactly generated Lie group, then by Corollary 5.5.14 we only have to look for a composition sequence of open subsets $(W_\nu)_{0 \leq \nu \leq \sigma}$ for $\widehat{N}$ such that the symmetrizer map is continuous on each $W_{\nu+1} \smallsetminus W_\nu$, since by Theorem 3.2.1, the type I'ness of the Mackey obstructions already follows from the type I'ness of $C^*(G)$. Recall from Remark 5.5.15 that in case where $\tilde{G}$ is a compactly generated Lie group, the answer to the question whether ω_χ is type I or not only depends on the structure of G/Σ_χ.

This becomes most satisfying if G/N is a vector group. Recall that the set of all vector subgroups with fixed dimension in a vector group V forms on open and compact subset of the set of all vector subgroups of V. This well known fact follows for instance from Lemma 5.5.18 (in fact it is well known that the set of k-dimensional vector subgroups of $\mathbb{R}^n$, equipped with the topology induced from $\mathfrak{K}(\mathbb{R}^n)$, is equal to the corresponding subspace of the Grassmannian manifold $\mathrm{Gr}(n,k)$). Moreover, if ω is a multiplier on V, then ω is type I and there exists a skew symmetric form $f : V \times V \to \mathbb{R}$ on V such that $\omega(s,t) = e^{if(s,t)}$ for all $s, t \in V$. It follows that the symmetry group Σ_ω is equal to the radical of f. But this implies that Σ_ω is a vector group of even codimension in V. Thus, it follows easily from Corollary 5.4.6 that, if G/N is a vector group, the symmetrizer map for $(C_0(\widehat{N}), G, \gamma^N, \tau^N)$ is continuous on a set $W \subseteq \widehat{N}$ if and only if the symmetrizers Σ_χ have constant codimension in G on W. In fact, we can use the following more general result.

THEOREM 6.3.3. *Suppose that (A, G, α, τ) is a separable abelian twisted system such that A has continuous trace and such that G acts trivially on $\widehat{A}$. Suppose further that $\tilde{G} = G/N_\tau$ is a vector group of dimension n. Then there exists a finite sequence of open sets*

$$\emptyset = W_0 \subseteq W_1 \subseteq \ldots W_k = \widehat{A},$$

with $k \leq \frac{n}{2} + 1$, such that the symmetrizer map Σ is continuous on $W_l \smallsetminus W_{l-1}$ for all $1 \leq l \leq k$. As a consequence, if we put $I_0 = \{0\}$ and $I_l = A_{W_l} \rtimes_{\alpha,\tau} G$ for $l > 0$, then each I_l is an ideal of $A \rtimes_{\alpha,\tau} G$ such that

$$\{0\} = I_0 \subseteq I_1 \subseteq \ldots \subseteq I_k = A \rtimes_{\alpha,\tau} G$$

and such that each quotient I_l/I_{l-1}, $1 \leq l \leq k$, is a C^*-algebra with continuous trace.

PROOF. For each $0 \leq l \leq \frac{n}{2} + 1$ let W_l denote the set of all $\rho \in \widehat{A}$ such that $\tilde{\Sigma}_\rho$ has dimension less or equal to $2l - 1$. Then it follows from Corollary 5.4.6 that each W_l is open in $\widehat{A}$. Since the symmetry groups have even codimension in G, it follows that

$$\emptyset = W_0 \subseteq W_1 \subseteq \ldots W_k = \widehat{A}$$

if k denotes the largest natural number which is less or equal to $\frac{n}{2} + 1$. If we define $I_l = A_{W_l} \rtimes_{\alpha,\tau} G$, then it follows that

$$\{0\} = I_0 \subseteq I_1 \subseteq \ldots \subseteq I_k = A \rtimes_{\alpha,\tau} G.$$

By construction, we have $2l - 3 < \dim \tilde{\Sigma}_\rho \leq 2l - 1$ for all $\rho \in W_l \setminus W_{l-1} = \widehat{A}_{W_l \setminus W_{l-1}}$. But since G/Σ_ρ has even dimension for all $\rho \in \widehat{A}$, this implies that the dimension of $\tilde{\Sigma}_\rho$ is constant for all $\rho \in W_l \setminus W_{l-1}$. Thus, the symmetrizer map is continuous on $W_l \setminus W_{l-1}$ and $I_l/I_{l-1} \cong A_{W_l \setminus W_{l-1}} \rtimes_{\alpha,\tau} G$ has continuous trace by Corollary 5.5.14. $\square$

As a direct consequence we obtain

COROLLARY 6.3.4. *Suppose that G is a second countable two-step nilpotent group with center N such that $\tilde{G} = G/N$ is a vector group of dimension n. Then there exists a chain*

$$\{0\} = I_0 \subseteq I_1 \subseteq \ldots \subseteq I_k = C^*(G)$$

of ideals in $C^*(G)$ such that $k \leq \frac{n}{2} + 1$ and I_l/I_{l-1} has continuous trace for all $1 \leq l \leq k$.

It might be interesting to see a concrete example for such a decomposition. The easiest example is given by the Heisenberg group H (compare with the remarks preceding Conjecture 5.5.12). Recall that $H = \mathbb{R}^3$ equipped with multiplication

$$(s,t,r)(s',t',r') = (s + s', t + t', r + r' + st').$$

The center N is equal to $(0,0,\mathbb{R})$, and we have $C^*(N) \cong C_0(\mathbb{R})$. If $r \in \mathbb{R}^* = \mathbb{R} \setminus \{0\}$, viewed as a representation of $C_0(\mathbb{R})$, then the symmetrizer of r for the system $(C_0(\mathbb{R}), H, \gamma^N, \tau^N)$ is equal to N, while the symmetrizer is all of H for $r = 0$. Thus, a composition sequence for $C^*(H)$ is given by

$$\{0\} = I_0 \subseteq I_1 \subseteq I_2 = C^*(H)$$

with $I_1 = C_0(\mathbb{R}^*) \rtimes_{\gamma^N, \tau^N} H$ (which is isomorphic to $C_0(\mathbb{R}^*, \mathcal{K}(L^2(\mathbb{R})))$) and $I_2/I_1 = C(\{0\}) \rtimes_{\gamma^N, \tau^N} H \cong C_0(\mathbb{R}^2)$. Of course, this example is well known, but it illustrates the procedure. In the following we present an example which is a little bit more complicated. Recall that all simply connected nilpotent groups with dimension less or equal to six are classified in [**45**], where additional interesting data for these groups are given.

Example 6.3.5. Let $G = G_{6,15}$ in Nielsen's list [**45**], that is $G = \mathbb{R}^6$ equipped with the multiplication

$$(s_1, s_2, s_3, s_4, s_5, s_6)(t_1, t_2, t_3, t_4, t_5, t_6)$$
$$= (s_1 + t_1 + s_6 t_4, s_2 + t_2 + s_5 t_4, s_3 + t_3 + s_6 t_5, s_4 + t_4, s_5 + t_5, s_6 + t_6).$$

Then G is two-step nilpotent with center $N = (\mathbb{R}, \mathbb{R}, \mathbb{R}, 0, 0, 0)$. Since $C^*(N) \cong C_0(\mathbb{R}^3)$ we may write $C^*(G) = C_0(\mathbb{R}^3) \rtimes_{\gamma^N, \tau^N} G$. Let $(r_1, r_2, r_3) \in \mathbb{R}^3$. If

$(r_1, r_2, r_3) \neq (0,0,0)$, then the symmetry group $\Sigma_{(r_1,r_2,r_3)}$ of (r_1, r_2, r_3) is given by

$$\Sigma_{(r_1,r_2,r_3)} = \{(s_1, s_2, s_3, tr_3, -tr_1, tr_2) \in G; s_1, s_2, s_3, t \in \mathbb{R}\},$$

while $\Sigma_{(0,0,0)} = G$. This follows very easily from the data given in [45].

Thus, if $(r_1, r_2, r_3) \neq (0,0,0)$, then the codimension of $\Sigma_{(r_1,r_2,r_3)}/N$ equals two and the symmetrizer map is continuous on $\Omega = \mathbb{R}^3 \setminus \{(0,0,0)\}$. It follows that $C_0(\Omega) \rtimes_{\gamma^N, \tau^N} G$ has continuous trace. Thus we get the composition sequence

$$\{0\} = I_0 \subseteq I_1 \subseteq I_2 = C^*(G)$$

with $I_1 = C_0(\Omega) \rtimes_{\gamma^N, \tau^N} G$ and $I_2/I_1 \cong C_0(\mathbb{R}^3)$.

Moreover, the dual space of I_1 is homeomorphic to $(C_0(\Omega) \rtimes_{\gamma^N, \tau^N} \Omega^\Sigma)^{\widehat{\ }}$ by Lemma 5.5.7. But the latter space is homeomorphic to $\Omega \times \mathbb{R}$ via the map

$$f : \Omega \times \mathbb{R} \to (C_0(\Omega) \rtimes_{\gamma^N, \tau^N} \Omega^\Sigma)^{\widehat{\ }}; (r_1, r_2, r_3, r_4) \to ((r_1, r_2, r_3), \chi_{(r_1,r_2,r_3,r_4)}),$$

where $\chi_{(r_1,r_2,r_3,r_4)} \in \widehat{\Sigma}_{(r_1,r_2,r_3)}$ is defined by

$$\chi_{(r_1,r_2,r_3,r_4)}(s_1, s_2, s_3, tr_3, -tr_1, tr_2) = e^{i(r_1 s_1 + r_2 s_2 + r_3 s_3 + r_4 t)}.$$

Thus we conclude that $\widehat{I_1}$ is homeomorphic to $(\mathbb{R}^3 \setminus \{(0,0,0)\}) \times \mathbb{R} \cong S^2 \times \mathbb{R}^2$, where S^2 denotes the two-sphere. Since third cohomology is trivial on S^2 (and hence also on $S^2 \times \mathbb{R}^2$), this implies that I_1 has a trivial Dixmier-Douady invariant. Thus, it follows from [9, Theorem 10.8.8] that $I_1 \cong C_0(S^2 \times \mathbb{R}^2, \mathcal{K})$.

Note that in the same way we could easily treat all other two-step nilpotent groups in Nielson's list, which are the groups $G_{5,2}, G_{6,16}, G_{6,17}$. We didn't take the five-dimensional group $G_{5,2}$ for the example above, since this group might be handled easier by writing it as a semidirect product $\mathbb{R}^4 \rtimes \mathbb{R}$ and using the well known results for transformation group C^*-algebras. Let us also remark that it is well known in the literature that connected nilpotent Lie groups always have a finite composition sequence of ideals with continuous trace subquotients [8, 51]. However, the composition sequence given in the example above is best possible, while the composition sequence given in [51] applied to our example has length 4. So we think that our methods provide new results even for simply connected two-step nilpotent groups.

Let us finally mention that a similar procedure as suggested above can be used for the study of group C^*-algebras of connected solvable Lie groups. If G is such a group, then $N = [G, G]$ is a connected nilpotent Lie group, and by [51, Theorem 4.3.1] we know that $C^*(N)$ has generalized continuous trace of finite length. By [17, Remark 2] we can therefore assume that there exists a finite composition sequence of G-invariant ideals

$$\{0\} = I_0 \subseteq I_1 \subseteq \ldots \subseteq I_m = C^*(N),$$

such that $A_l = I_l/I_{l-1}$ has continuous trace for all $1 \leq l \leq m$. Thus we obtain a composition sequence for $C^*(G)$ given by the ideals

$$\{0\} = J_0 \subseteq J_1 \subseteq \ldots \subseteq J_m = C^*(G),$$

with $J_l = I_l \rtimes_{\gamma^N, \tau^N} G$ and $J_l/J_{l-1} = A_l \rtimes_{\gamma^N, \tau^N} G$ for all $1 \leq l \leq m$, and we may apply our results to the systems $(A_l, G, \gamma^N, \tau^N)$.

Example 6.3.6. Let G be the so-called oscillater group, that is $G = H \rtimes \mathbb{R}$, where $\mathbb{R}$ acts on the Heisenberg group H as follows:

$$t(r_1, r_2, r_3) = (r_1 \cos t - r_2 \sin t, r_1 \sin t + r_2 \cos t, r_3),$$

for $t \in \mathbb{R}$ and $(r_1, r_2, r_3) \in H$. Since we have a semidirect product, we may write $C^*(G) = C^*(H) \rtimes_\alpha \mathbb{R}$. As we saw earlier, we have the following composition sequence for $C^*(H)$:

$$\{0\} = I_0 \subseteq I_1 \subseteq I_2 = C^*(H),$$

where $I_1 \cong C_0(\mathbb{R}^*, \mathcal{K})$, $\mathbb{R}^* = \mathbb{R} \setminus \{0\}$, and $I_2/I_1 \cong C_0(\mathbb{R}^2)$. It is clear that both ideals are invariant under the action of $\mathbb{R}$. Since $\mathbb{R}$ doesn't act on the center Z of H, we see that $\mathbb{R}$ acts trivially on $\widehat{I_1}$. Thus we conclude that $I_1 \rtimes_\alpha \mathbb{R}$ has continuous trace. In fact, using [**54**, Corollary 0.14 and Remark 0.15], it is easily seen that $I_1 \rtimes_\alpha \mathbb{R} \cong C_0(\mathbb{R}^* \times \mathbb{R}, \mathcal{K})$. Now let us consider the action on $I_2/I_1 \cong C_0(\mathbb{R}^2)$. Since $\mathbb{R}$ acts on $\mathbb{R}^2$ via rotation, we see that we have a constant stabilizer $2\pi\mathbb{Z}$ for all $(s_1, s_2) \neq (0, 0)$, while the stabilizer at $(0, 0)$ is $\mathbb{R}$. It is clear that $\mathbb{R}/2\pi\mathbb{Z}$ acts properly on $\mathbb{R}^2 \setminus \{(0, 0)\}$, since it is compact. Thus we see that $C_0(\mathbb{R}^2 \setminus \{(0, 0)\}) \rtimes_\alpha \mathbb{R}$ has continuous trace. In fact, it is not hard to see that $C_0(\mathbb{R}^2 \setminus \{(0, 0)\}) \rtimes_\alpha \mathbb{R}$ is isomorphic to $C_0(\mathbb{T} \times \mathbb{R}, \mathcal{K})$. Now let $\tilde{I}_1 = I_1$, $\tilde{I}_2 = \ker 1_H \subseteq C^*(H)$, the kernel of the trivial representation of H, and put $J_1 = \tilde{I}_1 \rtimes_\alpha \mathbb{R}$ and $J_2 = \tilde{I}_2 \rtimes_\alpha \mathbb{R}$. Then we have a composition sequence

$$\{0\} = J_0 \subseteq J_1 \subseteq J_2 \subseteq J_3 = C^*(G),$$

such that $J_1 \cong C_0(\mathbb{R}^* \times \mathbb{R}, \mathcal{K})$, $J_2/J_1 \cong C_0(\mathbb{T} \times \mathbb{R}, \mathcal{K})$ and $J_3/J_2 \cong C_0(\mathbb{R})$.

Note that all results used in the previous example have been available before we wrote this paper. This changes if we investigate more complicated groups. However, we chose this example since it demonstrates quite well the usefulness of the general approach.

Some concluding remarks

In the previous chapters we gave quite satisfactory results towards the description of abelian twisted systems (A, G, α, τ) with the property that A and $A \rtimes_{\alpha,\tau} G$ have continuous trace. However, besides the problems about the Hausdorff assumption in Theorem 5.4.3 and the correctness of Conjecture 5.5.12, there are still many open problems which should be investigated in future.

One of these problems is whether it is possible to give satisfactory results for computing the Dixmier-Douady class of $A \rtimes_{\alpha,\tau} G$ in terms of the Dixmier-Douady class of A and the action of G on A. The results about the connection between $\delta(A)$ and $\delta(A \rtimes_{\alpha,\tau} G)$ given in Theorems 5.3.2, 5.3.3 and 6.2.1 are far away from determining $\delta(A \rtimes_{\alpha,\tau} G)$. However, in some special situations there are some better results available (see for instance [19, 44, 54, 58, 59, 60]).

Another important problem is to describe non-abelian systems (A, G, α, τ) such that A and $A \rtimes_{\alpha,\tau} G$ have continuous trace. There are some results available for transformation groups (G, Ω) [32, 70, 14], but even in this situation there is no complete answer to the problem (however, in this case the problem is completely solved if all stabilizers are abelian or if G is compact or discrete [14]). If A is a non-commutative continuous-trace C^*-algebra, and if G acts freely and properly on $\widehat{A}$, then it is part of [54, Theorem 1.1] that $A \rtimes_\alpha G$ has continuous trace. So it would be interesting to know whether the converse is true if G is non-abelian. This can possibly be proved by using non-abelian duality (see for instance [53] for this theory). We also have some hope that at least the case where G is compact could be handled in the not to far future.

Finally, let us recall the conjecture given by Raeburn and Rosenberg in [54] (see also [65]) which says that every continuous trace subquotient of the group C^*-algebra of a connected nilpotent, or, more generally, exponential solvable Lie group has trivial Dixmier-Douady invariant. In Example 6.3.5 above, this follows from the simple fact that third cohomology is trivial on the dual spaces of the ideals of the composition series. In fact, using some results of this paper and a beautiful argument involving semi-simple Lie groups, Rosenberg was able to prove that his conjecture is true at least for two-step nilpotent groups [37, Theorem 3.4].

References

1. L. Auslander and C. C. Moore. *Unitary representations of solvable Lie groups.* Memoirs Amer. Math. Soc., **62** (1966).

2. L. W. Baggett, A. L. Carey, W. Moran, and A. Ramsay. *Non-monomial multiplier representations of abelian groups.* J. Funct. Anal., **97** (1991), 361–372.

3. L. W. Baggett and A. Kleppner. *Multiplier representations of abelian groups.* J. Funct. Anal., **14** (1973), 299–324.

4. L. G. Brown, P. Green, and M. A. Rieffel. *Stable isomorphism and strong Morita equivalence of* C^*-*algebras.* Pacific J. Math., **71** (1977), 349–363.

5. A. L. Carey and W. Moran. *Non-monomial unitary representations of nilpotent groups I: The abelian case.* Proc. London Math. Soc. (3), **46** (1983), 53–82.

6. F. Combes. *Crossed products and Morita equivalence.* Proc. London Math. Soc. (3), **49** (1984), 289–306.

7. J. Dixmier. *Traces sur les* C^*-*algèbres.* Ann. Inst. Fourier, **13** (1963), 219–262.

8. _____ *Sur le dual d'un groupe de Lie nilpotent.* Bull. Sci. Math., **90** (1966), 113–118.

9. _____ C^*-*algebras.* North-Holland, New York, 1977.

10. S. Echterhoff. *On induced covariant systems.* Proc. Amer. Math. Soc., **108** (1990), 703–706.

11. _____ *On maximal prime ideals in certain group* C^*-*algebras and crossed product algebras.* J. Operator Theory, **23** (1990), 317–338.

12. _____ *The primitive ideal space of twisted covariant systems with continuously varying stabilizers.* Math. Ann., **292** (1992), 59–84.

13. _____ *Regularizations of twisted covariant systems and crossed products with continuous trace.* J. Funct. Anal., **116** (1993), 277–314.

14. _____ *On transformation group* C^*-*algebras with continuous trace.* Trans. Amer. Math. Soc., **343** (1994), 117–133.

15. _____ *Morita equivalent actions and a new version of the Packer-Raeburn stabilization trick.* J. London Math. Soc. (2), **50** (1994), 170–186.

16. _____ *Duality of induction and restriction for abelian twisted covariant systems.* Math. Proc. Camb. Phil. Soc., **116** (1994), 301–315.

17. S. Echterhoff and E. Kaniuth. *Certain group extensions and twisted covariance algebras with generalized continuous trace.* In: Harmonic Analysis (Proceedings, 1987), Lecture Notes in Math., vol. 1359, pages 159–169, Springer-Verlag, Berlin, Heidelberg, New York, 1988.

18. S. Echterhoff and I. Raeburn. *The stabilization trick for coactions.* preprint.

19. S. Echterhoff and J. Rosenberg. *Fine structure of the Mackey machine for actions of abelian groups with constant Mackey obstruction. Pacific J. Math.* (to appear).

20. S. Echterhoff and D. P. Williams. *Crossed products whose primitive ideal spaces are generalized trivial $\hat{G}$-bundles.* Math. Ann. (to appear).

21. J. M. G. Fell. *A Hausdorff topology on the closed subsets of a locally compact non-Hausdorff space.* Proc. Amer. Math. Soc., **13** (1962), 472–476.

22. _____ *Weak containment and induced representations of groups.* Canad. J. Math., **14** (1962), 237–268.

23. _____ *Weak containment and Kronecker products of group representations.* Pacific J. Math., **13** (1963), 503–510.

24. _____ *Weak containment and induced representations of groups, II.* Trans. Amer. Math. Soc., **110** (1964), 424–447.

25. _____ *An extension of Mackey's method to Banach $*$-algebraic bundles.* Memoirs Amer. Math. Soc., **90** (1969).

26. A. Gleason. *Spaces with a compact Lie group of transformations.* Proc. Amer. Math. Soc., **1** (1950), 35–43.

27. J. Glimm. *Families of induced representations.* Pacific J. Math., **72** (1962), 885–911.

28. E. C. Gootman. *Abelian group actions on type I C^*-algebras.* In: Operator Algebras and their Connections with Topology and Ergodic Theory, vol. 1132 of Lecture Notes in Mathematics, pp. 152–169, Buşteni, Romania, 1983. Springer-Verlag.

29. E. C. Gootman and A. Lazar. *Crossed products of type I AF algebras by abelian groups.* Israel J. Math., **56** (1986), 267–279.

30. E. C. Gootman and D. Olesen. *Spectra of actions on type I C^*-algebras.* Math.-Scand., **47** (1980), 329–349.

31. E. C. Gootman and J. Rosenberg. *The structure of crossed product C^*-algebras: A proof of the generalized Effros-Hahn conjecture.* Invent. Math., **52** (1979), 283–298.

32. P. Green. C^*-*algebras of transformation groups with smooth orbit space.* Pacific J. Math., **72** (1977), 71–97.

33. _____ *The local structure of twisted covariance algebras.* Acta. Math., **140** (1978), 191–250.

34. K. C. Hannabuss. *Representations of nilpotent locally compact groups.* J. Funct. Anal., **34** (1974), 146–165.

35. S. Hurder, D. Olesen, I. Raeburn, and J. Rosenberg. *The Connes spectrum for actions of abelian groups on continuous-trace algebras.* Ergod. Th. & Dynam. Sys., **6** (1986), 541–560.

36. R.-Y. Lee. *On the C*-algebras of operator fields.* Indiana Univ. Math. J., **25** (1976), 303–314.

37. R. L. Lipsman and J. Rosenberg. *The behavior of Fourier transforms for nilpotent Lie groups.* preprint.

38. G. W. Mackey. *Imprimitivity for representations of locally compact groups, I.* Proc. Nat. Acad. Sci. U. S. A., **35** (1949), 537–545.

39. _____ *On induced representations of groups.* Amer. J. Math., **73** (1951), 576–592.

40. _____ *Induced representations of locally compact groups. I.* Ann. Math., **55** (1952), 101–139.

41. _____ *Unitary representations of group extensions. I.* Acta math., **99** (1958), 265–311.

42. C. C. Moore. *Group extensions and cohomology for locally compact groups. III.* Trans. Amer. Math. Soc., **221** (1976), 1–33.

43. _____ *Group extensions and cohomology for locally compact groups. IV.* Trans. Amer. Math. Soc., **221** (1976), 34–58.

44. P. S. Muhly and D. P. Williams. *Transformation group C*-algebras with continuous trace. II.* J. Operator Theory, **11** (1984), 109–124.

45. O. A. Nielsen. *Unitary representations and coadjoint orbits of low-dimensional nilpotent Lie groups.* Queen's Papers in Pure and Applied Mathematics No 63, Kingston 1983.

46. D. Olesen and I. Raeburn. *Pointwise unitary automorphism groups.* J. Funct. Anal., **93** (1990), 278–309.

47. J. A. Packer and I. Raeburn. *Twisted crossed products of C*-algebras.* Math. Proc. Camb. Phil. Soc., **106** (1989), 293–311.

48. R. S. Palais. *On the existence of slices for actions of non-compact Lie groups.* Ann. Math., **73** (1961), 295–323.

49. G. K. Pedersen. *A decomposition theorem for C*-algebras.* Math.-Scand., **22** (1968), 266–268.

50. _____ *C*-Algebras and their Automorphism Groups.* Academic Press, London, 1979.

51. N. V. Pedersen. *On the infinitesimal kernel of irreducible representations of nilpotent Lie groups.* Bull. Soc. math. France, **112** (1984), 423–467.

52. J. Phillips and I. Raeburn. *Crossed products by locally unitary automorphism groups and principal bundles.* J. Operator Theory, **11** (1984), 215–241.

53. I. Raeburn. *On crossed products by coactions and their representation theory.* Proc. London Math. Soc., **64** (1992), 625–652.

54. I. Raeburn and J. Rosenberg. *Crossed products of continuous-trace C*-algebras by smooth actions.* Trans. Amer. Math. Soc., **305** (1988), 1–45.

55. I. Raeburn and D. P. Williams. *Pull-backs of* C*-*algebras and crossed products by certain diagonal actions.* Trans. Amer. Math. Soc., **287** (1985), 755–777.

56. _____ *Crossed products by actions which are locally unitary on the stabilisers.* J. Funct. Anal., **81** (1988), 385–431.

57. _____ *Moore cohomology, principal bundles, and actions of groups on* C*-*algebras.* Indiana U. Math. J., **40** (1991), 707–740.

58. _____ *Topological invariants associated to the spectrum of crossed product* C*-*algebras.* J. Funct. Anal., **116** (1993), 245–276.

59. _____ *Dixmier-Douady classes for dynamical systems and crossed products.* Canad. J. Math., **45**(5), (1993), 1032–1066.

60. _____ *Equivariant cohomology and a Gysin sequence for principal bundles.* preprint.

61. J. Renault. *Reprśentations des produits croisés d'algèbres de groupoides.* J. Operator Theory., **18** (1987), 361–363.

62. M. A. Rieffel. *Induced representations of* C*-*algebras.* Adv. in Math., **13** (1974), 176–257.

63. _____ *Unitary representations of group extensions: an algebraic approach to the theory of Mackey and Blattner.* Adv. in Math. Supplementary Studies, **4** (1979), 43–81.

64. J. Rosenberg. *Some results on cohomology with Borel cochains, with applications to group actions on operator algebras.* Operator Theory: Advances and Applications, **17** (1986), 301–330.

65. _____ C*-*algebras and Mackey's theory of group representations.* In: C*-algebras: 1943–1993, a Fifty-Year Celebration (R. S. Doran, ed.), Contemp. Math., vol. 167, Amer. Math. Soc., Providence, RI, 1994.

66. _____ *Continuous-trace algebras from the bundle-theoretic point of view.* J. Aust. Math. Soc. (Series A), **47** (1989), 368–381.

67. J.-L. Sauvageot. *Ideaux primitifs de certain produits croises.* Math. Ann., **231** (1977), 61–76.

68. I. Schochetman. *The dual topology of certain group extensions.* Adv. in Math., **35** (1980), 113–128.

69. M. Takesaki. *Covariant representations of* C*-*algebras and their locally compact automorphism groups.* Acta Math., **119** (1967), 273–303.

70. D. P. Williams. *Transformation group* C*-*algebras with continuous trace.* J. Funct. Anal., **41** (1981), 40–76.

71. _____ *Transformation group* C*-*algebras with Hausdorff spectrum.* Illinois J. Math., **26** (1982), 317–321.

Editorial Information

To be published in the *Memoirs*, a paper must be correct, new, nontrivial, and significant. Further, it must be well written and of interest to a substantial number of mathematicians. Piecemeal results, such as an inconclusive step toward an unproved major theorem or a minor variation on a known result, are in general not acceptable for publication. *Transactions* Editors shall solicit and encourage publication of worthy papers. Papers appearing in *Memoirs* are generally longer than those appearing in *Transactions* with which it shares an editorial committee.

As of May 31, 1996, the backlog for this journal was approximately 7 volumes. This estimate is the result of dividing the number of manuscripts for this journal in the Providence office that have not yet gone to the printer on the above date by the average number of monographs per volume over the previous twelve months, reduced by the number of issues published in four months (the time necessary for preparing an issue for the printer). (There are 6 volumes per year, each containing at least 4 numbers.)

A Copyright Transfer Agreement is required before a paper will be published in this journal. By submitting a paper to this journal, authors certify that the manuscript has not been submitted to nor is it under consideration for publication by another journal, conference proceedings, or similar publication.

Information for Authors and Editors

Memoirs are printed by photo-offset from camera copy fully prepared by the author. This means that the finished book will look exactly like the copy submitted.

The paper must contain a *descriptive title* and an *abstract* that summarizes the article in language suitable for workers in the general field (algebra, analysis, etc.). The *descriptive title* should be short, but informative; useless or vague phrases such as "some remarks about" or "concerning" should be avoided. The *abstract* should be at least one complete sentence, and at most 300 words. Included with the footnotes to the paper, there should be the 1991 *Mathematics Subject Classification* representing the primary and secondary subjects of the article. This may be followed by a list of *key words and phrases* describing the subject matter of the article and taken from it. A list of the numbers may be found in the annual index of *Mathematical Reviews*, published with the December issue starting in 1990, as well as from the electronic service e-MATH [**telnet e-MATH.ams.org** (or **telnet 130.44.1.100**). Login and password are **e-math**]. For journal abbreviations used in bibliographies, see the list of serials in the latest *Mathematical Reviews* annual index. When the manuscript is submitted, authors should supply the editor with electronic addresses if available. These will be printed after the postal address at the end of each article.

Electronically prepared papers. The AMS encourages submission of electronically prepared papers in $\mathcal{A}_{\mathcal{M}}\mathcal{S}$-TEX or $\mathcal{A}_{\mathcal{M}}\mathcal{S}$-LATEX. The Society has prepared author packages for each AMS publication. Author packages include instructions for preparing electronic papers, the *AMS Author Handbook*, samples, and a style file that generates the particular design specifications of that publication series for both $\mathcal{A}_{\mathcal{M}}\mathcal{S}$-TEX and $\mathcal{A}_{\mathcal{M}}\mathcal{S}$-LATEX.

Authors with FTP access may retrieve an author package from the Society's Internet node `e-MATH.ams.org` (130.44.1.100). For those without FTP access, the author package can be obtained free of charge by sending e-mail to **pub@math.ams.org** (Internet) or from the Publication Division, American Mathematical Society, P.O. Box 6248, Providence, RI 02940-6248. When requesting an author package, please specify $\mathcal{AMS}$-TEX or $\mathcal{AMS}$-LATEX, Macintosh or IBM (3.5) format, and the publication in which your paper will appear. Please be sure to include your complete mailing address.

Submission of electronic files. At the time of submission, the source file(s) should be sent to the Providence office (this includes any TEX source file, any graphics files, and the DVI or PostScript file).

Before sending the source file, be sure you have proofread your paper carefully. The files you send must be the EXACT files used to generate the proof copy that was accepted for publication. For all publications, authors are required to send a printed copy of their paper, which exactly matches the copy approved for publication, along with any graphics that will appear in the paper.

TEX files may be submitted by email, FTP, or on diskette. The DVI file(s) and PostScript files should be submitted only by FTP or on diskette unless they are encoded properly to submit through e-mail. (DVI files are binary and PostScript files tend to be very large.)

Files sent by electronic mail should be addressed to the Internet address **pub-submit@math.ams.org**. The subject line of the message should include the publication code to identify it as a Memoir. TEX source files, DVI files, and PostScript files can be transferred over the Internet by FTP to the Internet node **e-math.ams.org** (130.44.1.100).

Electronic graphics. Figures may be submitted to the AMS in an electronic format. The AMS recommends that graphics created electronically be saved in Encapsulated PostScript (EPS) format. This includes graphics originated via a graphics application as well as scanned photographs or other computer-generated images.

If the graphics package used does not support EPS output, the graphics file should be saved in one of the standard graphics formats—such as TIFF, PICT, GIF, etc.—rather than in an application-dependent format. Graphics files submitted in an application-dependent format are not likely to be used. No matter what method was used to produce the graphic, it is necessary to provide a paper copy to the AMS.

Authors using graphics packages for the creation of electronic art should also avoid the use of any lines thinner than 0.5 points in width. Many graphics packages allow the user to specify a "hairline" for a very thin line. Hairlines often look acceptable when proofed on a typical laser printer. However, when produced on a high-resolution laser imagesetter, hairlines become nearly invisible and will be lost entirely in the final printing process.

Screens should be set to values between 15% and 85%. Screens which fall outside of this range are too light or too dark to print correctly.

Any inquiries concerning a paper that has been accepted for publication should be sent directly to the Editorial Department, American Mathematical Society, P. O. Box 6248, Providence, RI 02940-6248.

Selected Titles in This Series

(*Continued from the front of this publication*)

(See the AMS catalog for earlier titles)